Beasts of the Sky

Beasts of the Sky: Strange Sightings in the Stratosphere

Edited by Jon Hackett, Damian O'Byrne and Seán Harrington

British Library Cataloguing in Publication Data

Beasts of the Sky: Strange Sightings in the Stratosphere

A catalogue entry for this book is available from the British Library

ISBN
Paperback: 9780 86196 754 4;
Electronic: 9780 86196 991 3 (EPUB);
Electronic: 9780 86196 992 0 (EPDF)

Published by
John Libbey Publishing, 205 Crescent Road, East Barnet, Herts EN4 8SB,
United Kingdom
e-mail: johnlibbeypublishing@gmail.com   web site: www.johnlibbey.com

Distributed worldwide by **Indiana University Press**,
Herman B Wells Library – 350, 1320 E. 10th St., Bloomington, IN 47405, USA.
www.iupress.indiana.edu

Contents

Acknowledgements

The editors would like to thank our long-suffering illustrator Rupert Norfolk, for all his talent and effort across all three volumes of the *Beasts* series. We would also like to thank our excellent graphic designer and talented colleague – Lee Brooks, for his work in designing the posters and book covers for each conference and publication. Last, but certainly not least, we would like to thank our publisher John Libbey for his support and hard work in putting these books in print and distribution.

Introduction

Beasts of the Sky

O ver the course of three conferences and several publications, we the editors have been privileged to work with a great many talented academics and writers. In each of our collections, we have endeavored to explore a curious facet of popular culture – the ways in which media represents the natural world, with a particular focus on the dark, malevolent or monstrous representations of specific geographies. Thus far we have plunged the depths of the deepest oceans (*Beasts of the Deep: Sea Creatures and Popular Culture*, 2017) and wandered through the darkest forests and woods (*Beasts of the Forest: Denizens of the Dark Woods*, 2019). Our previous volumes followed a discursive trajectory that examined monstrous apparitions that encapsulated culturally-expressed anxieties associated with the specific geographies – looking at representations of the monstrous and their contemporary cultural resonance. The final book in our little collection, casts our eyes upwards, to the great infinity of the sky, stratosphere, and beyond to outer space. The papers here are the product of a conference organised at St. Mary's University, Twickenham in the Summer of 2018. Our publication of this volume has been delayed, and at several points derailed, by the global Covid-19 pandemic – and indeed its impact and after-effects are still felt by many of us. Those readers who share our post-pandemic-fatigue, will be relieved to hear that Covid-19 is one airborne horror that we will not be unpacking herein. Instead we look beyond the world of our earthly toils, and turn our eyes skyward to the horrors of the sky and the great cosmic expanse. These are not the bright summer skies of a child's doodling – big happy sun and fluffy clouds – these are darker skies, looming overhead, home to UFOs, dragons, missiles and even the great vacuum of space – in which no-one will hear us scream. Perhaps the sky becomes a threatening location precisely because it looms omnipotent and omnipresent – to all but the most committed troglodytes and bunker-enthusiasts.

The sky and stars have functioned as a wondrous canvas for human imaginings – potentially since the dawn of imagination itself. Our earliest temples and sites of ritual were constructed as monuments to the relationship between the earth, the stars, moon and sun. Ireland's Newgrange passage tomb (estimated to have been built 3200 B.C.), Britain's Stonehenge (3000

B.C.) or even the pyramids at Giza (2500 B.C.), all stand as millenia-aged monuments to our fascination with the sky and its relationship with harvest, mortality and the afterlife. The neolithic peoples of Britain and Ireland in particular, built their great stone monuments to align with and chart the changing topography of the sun, moon and stars – to map and predict the changing seasons. The significance of seasonal change to settled, newly-agricultural communities cannot be overstated – a successful growing season and its harvest would have meant life or death to entire communities and cultures. From these early moments in social and cultural practice, the sky and stars became etched into the core concerns of our civilisations. The marked significance of the sky can be seen throughout our cultural progression – from specific theologies, for whom the sky becomes the location of heaven, to the objectives of the project of modernity, encapsulated in the 20th century 'space race'.

As we look to the skies in awe and wonder, we also see possible annihilation in the collapse of boundary between the earthly and the infinite – whether it be brought about by divine intervention, apocalypse or alien invasion. This calls to mind the adventures of Asterix and Obelix – ever frightened of an end of days, in which 'the sky will fall on our heads', until finally in *Asterix and the Falling Sky* (2005) aliens arrive in ancient Gaul – as harbingers of their feared apocalypse.

Back down here on earth – the following collection sees a series of papers that attempt to bite off and identify cultural apparitions of our relationship with the sky, and what strange sightings abound in media and popular culture. Whether it be aliens, sky-bison or missiles – the skies we discuss are full of all kinds of threatening apparitions. Whether real or imaginary, they allude to the great omnipotence and indifference of the above.

Our collection begins with Part 1: 'Filming Scary Skies', here we begin with two medium-specific papers looking at the representation of the sky or space as a threatening location. In our first chapter, Rachel Steward discusses the films depicting outer-space since the 1960s – suggesting that there has been a new 'presence' of the skies of our planetary age. In her analysis of *Gravity*, which along with *Oblivion* and *Elysium* form a trilogy of mainstream, near-Earth science fiction films all released in 2013. While all three films are analysed, it is *Gravity*'s planetary depiction that is the most remarkable. With critically acclaimed cinematography and visual effects – even the film's protagonists enjoy the setting. 'Can't beat the view… terrific' says Lieutenant Kowalski (Clooney) looking back at a luminous blue planet. How do we, the viewers, many of us children of the Apollo era feel about the 'terrific' view? What sentiments do these cinematic images of planet Earth floating in our space age skies evoke in us today? Read on and find out!

This is followed in Chapter 2 in which Sara Khalili conducts a contextual study of *A Short Vision*, an animated short by Peter and Joan Foldes produced in 1956 in Britain. Her aim, to understand how a few contextual components might reveal, create and/or may change the meaning of the film. To do such, she draws from Klinger (1997) – and discusses factors related to reception-studies as a framework, in order to briefly examine how the film was received upon release, as well as in recent decades. She explores how *A Short Vision* morphs into multiple visions, or an implicit layer or theme of "atomic war" explicates boldly, based on the contemporaneous events, including film distribution, exhibition, cross-cultural reception, plus biased writings and reports about the film with particular purposes.

Arthur C. Clarke famously said of space – "Two possibilities exist: either we are alone in the Universe or we are not. Both are equally terrifying." – in keeping with this sentiment Part 2 of our collection is entitled 'Aliens: Allies and Adversaries'! Here we continue the analyses of narrative film and their representations of the potential 'occupants' of outer space. James Williamson with his chapter 'Archetype as History in Hollywood Science Fiction of the 1950s', in which he discusses the interplanetary peacekeepers of *The Day the Earth Stood Still* (1951), and the genocidal adversaries of *The War of the Worlds* (1953). Both films reference the threat of nuclear conflict and planetary annihilation – a topic all to prescient of 2024 as the Doomsday Clock ticks ever closer to midnight. These films are analysed both as mythologised narratives – the approach taken by Jerome F. Shapiro in *Atomic Bomb Cinema*, and also as specific reactions to their contemporary context, emphasised by Peter Biskind in *Seeing is Believing*. As a drama rather than a special-effects spectacular, the themes of Robert Wise's *The Day the Earth Stood Still* are foregrounded in the relationships of its alien protagonist Klaatu, whose messianic qualities were noted by Hugh Ruppersberg. Religious and mythic archetypes are applied towards the film's secular, political address, which will be discussed in response to criticism of the film's message by Mark Jancovich.

Williamson's chapter is then followed by James Keys, who takes two well-documented current trends in the scholarly analysis of science fiction film. The first is genre criticism, the virtue of which, according to Annette Kuhn in her seminal book-length study *Alien Zone: Cultural Theory and Contemporary Science Fiction Cinema* (1990), is an appreciation of cinema in its totality as a social ritual, in which the spectator is situated as a cog in the machine of the cinematic apparatus. The second is sketched out in *Liquid Space: Science Fiction Film and Television in the Digital Age* (2017), in which Sean Redmond uses eye-tracking technology and cognitive research to explore the ways in which the spectator encounters spectacular moments, fueled by the development of digital technologies, in science fiction film. To exemplify his discussion, Keyes engages in an analysis of Denis Villeneuve's *Arrival* (2016) to integrate these lines of enquiry and move past them, using the influential work of genetic epistemologist Jean Piaget to argue that *Arrival* conceptualises cinematic grammar in science fiction film as both liberating and limiting.

Part 3 of our collection, 'Animating Sky-Borne Beasts' interrogates media specificity through a discussion of animation and digital games in their representations of sky and space-borne beasts. For his chapter, Francis M. Agnoli develops a twofold model for conceptual art and the rendering of 'Sky Bison' and 'Dragons' in *Avatar: The Last Airbender* and *The Legend of Korra*. First, he examines the discourse surrounding the development, design, and animation of these iconic characters. Both as hybridisations of transnational and transcultural inspirations and references, with the sky bison imitating Studio Ghibli designs – most notably *My Neighbor Totoro* (Miyazaki, 1988) – and the dragons blending Chinese and European characterisations. By looking at the process of characterization and conceptualisation, this chapter will reveal how the creators and crew articulated their navigation of these various transcultural elements. By comparing *Avatar* and *Korra*, this chapter will also discuss how fictional 'technological innovations' – especially in air travel – rendered these creatures as less special and awe-inspiring narratively, which mirrors the spectators own experience of these sky-borne beasts.

This is followed by a chapter from Chunning (Maggie) Guo, who discusses pioneering generative animation artists, whose transformative experiments see narrative structures morphed into digital evolution. The first generation of these animation works explore more about the mysterious creatures in ocean, from tracing models of marine organisms (*Aquarelles*, Tom De Witt) to 3D construction of digital creatures (*Growth Model*, Yoichiro Kawaguchi), generative animation is argued to have given more freedom to discuss the question of 'evolution' of form, and to have provided an opportunity to balance the debate of creationism and evolution. This is argued to also expand a large space for imagination of the diversity of an ecosystem in outer space.

The discussion of generative art is partly supported by the theory of 'digital genetic systems'. Guo suggests that through interactive behavior with audiences, these digital creatures from ocean to outer space are experiencing an adventure of re-evolution. Here the game *Spore*, created by Will Wright, could be regarded as a testament of re-evolution, in which players themselves create an entire universe from a single-celled creature – beginning life in water and eventually moving to inhabit outer space. Through vital abstract visual principles, *Spore* presented a process like evolution, which this chapter describes as 're-evolution'.

Our collection concludes with Part 4: 'Writing the Skies'. This final part looks at the representation of flying creatures in literature, and begins with Damian O'Byrne's chapter 'Fell Beasts and fell beasts: The Making of a Monster in J.R.R. Tolkien's The Lord of the Rings'. Here O'Byrne looks at the semiotic function and representational practice belying 'flying creatures' in Tolkien's work, in which there is a tendency to avoid creatures that are entirely 'fantastic' or without established 'mythical' precedent. There are no 'hybrids' or 'chimera' in *The Lord of the Rings* world, but rather creatures that connect to some existing or archaic understanding in the readership (or indeed – viewership of the Peter Jackson films). 'Fell Beasts' are the monstrous winged mounts of the Nazgûl – one of the many 'big bads' of the book series – these creatures are argued to encapsulate this tendency, which is used to maximise their horrific effect. Their form is alluded to and illusive, left to startling reveal and effect!

The final chapter offers readers something a little different – 'Buffalo Wings of Desire' is a short story by Jay Russell, who offers our readers some fictional respite after three volumes of monstrous geographies. Featuring a playful mixing of genre and iconographies (some less iconic than others) Russell takes us on another day in the life of his most recognisable literary creation – Marty Burns, supernatural detective. Marty navigates the world of high-flying beasties – dragons, demons and even some fellow named Daedalus ...

Bibliography

Hackett, Jon & Harrington, Seán (Eds.) (2017) *Beasts of the Deep: Sea Creatures and Popular Culture*, London, John Libbey Publishing

Hackett, Jon & Harrington, Seán (Eds.) (2019) *Beasts of the Forest: Denizens of the Dark Woods*, London, John Libbey Publishing

Uderzo, Albert (2005) *Asterix and the Falling Sky*, Sphere Publishing: London

Part 1

Filming Scary Skies

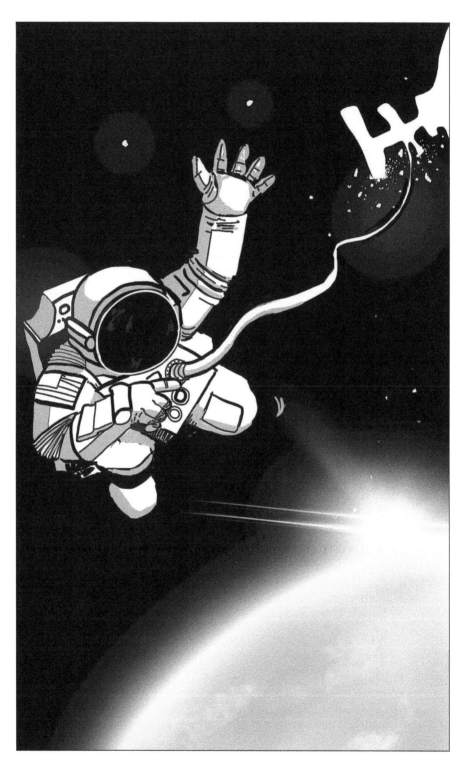

Chapter 1

Gravity's Earth

Rachel Steward

In 1948 the astronomer Fred Hoyle wrote: 'once a photograph of the Earth, taken from outside, is available, we shall, in an emotional sense, acquire an additional dimension'. He went on to imagine that at this time 'a new idea as powerful as any in history will be let loose' (Hoyle as quoted in Poole, 2014, p. 213). Hoyle was not alone in anticipating the powerful impact of the first views of our planetary home and, often citing from Hoyle, a great deal has been written since about the way photographs such as *The Blue Marble* (1972) led to the awakening of our 'planetary consciousness' (Sagan, 1994, p. 171). Over half a century after these first spellbinding views mesmerised the world, contemporary visual culture is still saturated with images of our planetary home.

In this chapter I focus on one such contemporary depiction, that created by Alfonso Cuarón in his award-winning film, *Gravity* (2013). Exploring *Gravity*'s portrait of Earth in light of the science-fiction cinematography that has come before it and also in relation to the broader visual culture from which Cuarón's imagery is drawn, I argue that *Gravity*'s Earth marks a radical shift in the history of cinematic landscape but, more than this, it draws our attention to changes in the way in which we depict our planetary home within mainstream culture today.

Context

Given this publication's theme of the sky and its beasts I wish briefly to explain what might be evident by now, why this chapter takes the Earth as its subject. The pleasure in being contrary need not be entirely discounted but the reasoning draws mainly from the fact that within our contemporary space age the conventional relationships between up and down, earth and sky are, as Paul Virilio so succinctly puts it, 'Arse over heels... Soon we will have to learn to fly, to swim in the ether' (Virilio, 1997, p. 3). When our astronauts and cosmonauts ruptured the blue sky travelling into the coal-black abyss of outer space everything changed. Things were no longer where they used to be. One such thing was our planetary home.

We have all experienced the strange sensation of seeing our planet suspended out in the infinite, black expanse of sky with shots of the Earth, taken from outside, reproduced in mainstream culture since the historic live broadcasts from Apollo 8 (1968). As historian Robert Poole recounts:

> The crew's second television transmission, on Monday 23 December, provided the first recognisable view of the whole Earth... 'There she is, floating in space!' exclaimed the newscaster Walter Cronkite. 'You are looking at yourselves from 180,000 miles out in space,' announced (Astronaut) Anders (Poole, 2008, pp. 21–22).

This space-age viewpoint has given our planetary home a new autonomy, for while we understand Earth as the steady ground upon which we stand, we have also witnessed it as a spectral, living presence floating in the expanse of the sky. It is this spectral Earth moving out there among the stars that is the subject of this chapter.

Gravity's Frame

Nowhere in contemporary culture is Earth, viewed from outside, more vividly portrayed than in Alfonso Cuarón's *Gravity*. Our planetary home is a powerful presence on the screen throughout the film. There are exceptions but from the opening sequence in which a luminous, blue-bodied planet appears within the black void of the screen, right through to the final re-entry and splash down, Earth plays a key role. This is not something we have seen before in cinema.

There are certainly many powerful planetary presences captured on film: Stanley Kubrick's planetary alignment in the opening and closing sequences of *2001: A Space Odyssey* (1968) remains in our brains like an afterimage on the retina; the Sun in Danny Boyle's *Sunshine* (2007) does the same, albeit in a more literal way; and arguably the Death Star haunts *Star Wars* (George Lucas, 1977) as some kind of tech-planet of doom. However, *Gravity*'s portrayal of Earth is different.

Cuarón describes Earth as 'a constant presence' and 'an ever-present character in the background' in the supplementary feature about the film, *Behind the Scenes*, 2014. In so speaking he is being quite literal. Earth is not depicted as part of a vista glimpsed from a spaceship portal or a viewing platform, but rather as the landscape through which the protagonists move. As Cuarón explains, the majority of science-fiction or space adventure films include only a short set piece actually outside the mothership, so for example 'there's a mission where they go outside the ship in a three-minute sequence with a lot of cuts' (Cuarón cited in Pond, 2013).

What is ground-breaking about *Gravity* is that for over half the film we watch the free-floating, extravehicular activity of the astronauts; the bulk of the action of the film is set 'outside'. First, we see three figures floating around the exterior of the Hubble telescope and the space shuttle Explorer. Then, we watch two tethered figures travel across the sky; propelled by one depleted thruster pack they make their way to the International Space

Station. In the final part of the film we follow the journey of the sole survivor, Dr Stone (Sandra Bullock), as she tries to find a way to get home. While in this final section Stone goes inside a spaceship for the first time, there are long, fraught sequences in which she moves around the exterior of the damaged Soyuz module and then the Chinese Space Station, Tiangong.

The impact of setting the bulk of the action directly within the landscape outstrips anything that could be achieved with dialogue. The power lies both in the length of time the audience is immersed within this beautiful spacescape, but also in the sheer scale of the Earth that they encounter there. What floats within *Gravity*'s cinematic frame for over ninety minutes is not the minuscule egg Michael Collins observed from the window of Apollo 11; it is not the blue marble Tom Hanks covers with his thumb in *Apollo 13* (Ron Howard, 1995); and it is not Voyager's pale, blue dot. *Gravity*'s Earth is enormous.

Cuarón's film is set at the relatively small distance of '600 km above planet Earth' (*Gravity*, 2013) and, in the words of Commander Lieutenant Matt Kowalski (George Clooney), the view from this vantage point is 'terrific'. At times all we see within the cinematic frame is the skin of our planet stretched out before us. We can identify landmass and seascape, cloud formations and mountain ranges. Cities glimmer like gossamer circuitry and sunlight turns oceans into molten pools. When the horizon and the black abyss beyond float back into view they are separated by the Earth's atmospheric limb, a local scattering of blue light, an ethereal halo the like of which adorns the painted heads of medieval saints and cherubs. Cuarón never draws back to show us the whole Earth and thus he denies us the pleasure of surveying a perfect, planetary object, a blue marble in all its completeness. Instead we remain mesmerisingly close to Earth's enormous, glowing form. At this distance all sense of scale seems to vanish and, dizzyingly bewitched, we watch in a kind of fetishised fascination as it looms large before us upon the screen. Is this Michelangelo Antonioni's *Blow-Up* (1966)? Or perhaps Grace Kelly's mouth, so close within Hitchcock's frame in *Rear Window* (1954) that we are transformed into adventurers starting out on a *Fantastic Voyage* (1966)?

Cuarón does not allow us to be swallowed, or drawn ever closer until all we can see is abstraction. Instead, with the exception of the final re-entry scene, he keeps us at the constant distance of 600 km 'up'. This is near-Earth or low-Earth orbit. In the present-day reality, that emulated in *Gravity*, this is where we find the International Space Station, the Hubble telescope and a growing collection of man-made debris encircling our planet. At this distance the pull of the Earth exerts a micro-gravitational force keeping these objects close. Yet, despite the draw of Earth, astronauts still experience weightlessness. It is the cinematic simulation of this weightlessness, of a world in which there isn't an 'up', that presents so many challenges for the filmmaker.

In the 1960s (for *2001*) Kubrick used scaffold and wires to create the cut-away sequences in which we watch Frank float and spin in the immensity of a voided sky, and the centrifugal force on board the Discovery was shot within a rotating, forty-foot drum. *Sunshine* features a short, external scene but does not attempt to simulate the weightless movement of zero or microgravity. Ron Howard made the ground-breaking decision to shoot key scenes of *Apollo 13* in the microgravity of parabolic flight. This is a specific arc of flight within the Earth's atmosphere during which the body experiences weightlessness for around 25 seconds before crashing back to the floor of the plane. By the end of shooting the crew and cast had taken 612 flights and experienced nearly four hours of weightlessness in 25-second sections. As a result, the film describes the vulnerability of three men sitting in a tin can (far above the world) by way of short shots and a claustrophobic tension.

Cuarón knew he wanted something different and from early conception designed the film 'around these extremely long, these very fluid twelve-minute, five-minute or three-minute shots' (*Behind the Scenes*, 2014). This is Tim Webber, visual effects supervisor, recalling early conversations with Cuarón in which they discussed the director's wish to deploy his signature long shot or tracking shot, seen in *Y tu mamá también* (2001) and *Children of Men* (2006), and the complex technical challenge this posed. The award-winning solutions resulted in a cinematic aesthetic that seamlessly simulates movement in microgravity regardless of whether we are watching a graceful, orbiting waltz of weightless or a frenetic mosh pit of collision. As film critic Ivan Radford summarises, 'the film's visceral simplicity was born out of one of the most complex productions in modern cinema history' (Radford, 2014).

The challenge of making a film set almost completely in microgravity is not only technical but also behavioural. The instinct to show things as they appear according to the gravitational conditions on the Earth's surface and the deeply embedded conventions of image-making that have arisen from these conditions needs to be unpicked. In simple terms, the instinct to make things appear 'normal' needs to be suppressed.

One of the most familiar examples of this instinct can be found in the reorientation of the photograph *Earthrise* (1968) which, as published by NASA, looks like a 'normal' landscape. The orientation shows the solid ground of the moon occupying the lower part of the image and the distanced celestial orb of the Earth rising upwards into the jet-black skies above. This orientation echoes what we have come to understand as landscape, whether that seen in the paintings of Caspar David Friedrich, the photographs of Ansel Adams or the spacescapes of Chesley Bonestell. Yet as the Apollo flight journal details, Bill Anders, who took the photograph, suggested:

> an alternative presentation, with the horizon running vertically, [which] represents how he saw this image. They were orbiting around the Moon's equator and, with north being to the top, Earth came out from behind a vertical horizon (O'Brien and Woods, 2004).

There are of course other famous examples of this normalising tendency. Carl Sagan (1994) writes of the colouring of the Martian sky to a default blue as this was what looked 'right', and Elizabeth Kessler (2012) traces the influence of frontier and romantic landscape painting on the composition, orientation and colouring of NASA's Hubble imagery. The influence of conventional landscape can also be tracked more broadly within NASA's archive (2019), raising interesting questions as to the degree to which NASA has normalised our engagement with the radical landscapes of outer space.

Cuarón was determined from the outset to unpick the auto-corrective tendencies innate with this genre. The director concedes: 'It took a lot of education for the animators to fully grasp that the usual laws of cause and effect don't apply' (*Behind the Scenes*, 2014). But it is a hard thing to try to unlearn. According to *Gravity*'s producer David Heyman, Cuarón constantly checked for mistakes. Heyman recounts one costly delay caused by Cuarón spotting that the Space Shuttle was orientated upwards for the duration of a shot (*Behind the Scenes*, 2014). The fact that the re-rendering of this sequence added two-and-a-half months to the film's already lengthy post-production process is indicative of the director's dedication to unlearn and unpick the norms of terra firma image-making.

This unpicking went beyond rethinking how things should appear within a frame but involved rethinking the idea of the frame itself. Within conventional film-making you establish a bottom edge or what is called a ground or floor within your frame. This fixed orientation informs the composition of each shot and in effect holds the background still while things float and move around within it (*Behind the Scenes*, 2014). Jerome Agel observes that this is what we experience watching Kubrick's centrifugal shots in *2001*, for while the Ferris wheel creates the impression that the astronauts move through 360 degrees, the frame in which the radical orientation occurs remains static (Agel, 1970).

In the real world of microgravity everything moves. There is no ground. 'Arse over heels… Soon we will have to learn to fly, to swim in the ether'. To capture this absolute movement, *Gravity*'s production team allowed the camera itself to appear as if floating. The result is that while *Gravity* relies heavily on NASA's archive and therefore inevitably reflects aspects of NASA's epic visual narrative, it also creates something new, an experience akin to that described by Nietzsche's madman: 'Backward, sideward, forward, in all directions? Is there still any up or down? Are we not straying as if through an infinite nothing?' (Nietzsche, 1974, p. 181). *Gravity* shows us a reality in which everything moves in all directions; this includes the camera and the background against which the film is set. We watch as planet Earth, large and blued, hovers to the left of the screen, and in the next moment it can be seen bottom right. This movement does not occur as one sequence jump cuts to another, but the movement appears to belong to the Earth itself, which glides across the screen. Its movement and size is perhaps less like Nietzsche's errant football and more like a majestic whale – large, fast and alive.

Gravity's depiction of 'Earth as a live, living entity' (Sandra Bullock in *Behind the Scenes*, 2014) moving through our space age skies is 100 percent computer-generated. While NASA's archives were referenced for accuracy, the artwork was all original. As Kyle McCullock, compositing supervisor, comments:

> we have a whole team of really gifted map-painters, every cloud in every shot was a hand-painted asset, and we would paint... and then sort of lay it out upon the Earth to see what we were going to see in the shot, and then those paintings would be changed into volumes, and then those were handed off to compositors who would sit and work them up until it really felt like the real thing (*Behind the Scenes*, 2014).

The detail of the art-working is evident as we watch the changing textures and tones of the living Earth. The oceans have shallows of translucent turquoise at the sandy coastlines and the exposed bones of mountain ranges cast shadows and herd clouds. Preparing the artwork was not only about getting 'it all physically accurate, we had to make sure it was beautiful' (Kyle McCullock in *Behind the Scenes*, 2014). The transformation from an immaculate copy of satellite surveillance data into an emotive, spellbinding landscape required the key ingredient of perfect light. This was a challenge, not only because the 'balance of light from Earth changes all the time' (Cuarón in *Behind the Scenes*, 2014), but because this nuance of moving colour and tone needed to be mapped across the composited digital and live footage.

The seamless way in which this was achieved within the film was a result of a collaboration between cinematographer Emmanuel 'Chivo' Lubezki and VFX supervisor Tim Webber. Together they designed a ten metre, cubed light box with walls made up of almost two million LED lights. The live footage, which in the zero-gravity sequences focused on the actors' faces, was filmed inside the light box. During each sequence the CGI pre-production material was screened on the LED walls allowing the actors to feel immersed in their surroundings but also allowing the colour and light created in the digital render to be captured in the live footage. This technique resulted in the CGI-ed Earth being transformed into what Tim Webber described as the main light source for the film (*Behind the Scenes*, 2014). To grasp the impact of this set-up, imagine a reverse aquarium, the actors inside a glass-walled cube and the Earth a magnificent whale gliding around them. As this enormous creature moves, diffused and direct light bounces off its smooth skin colouring everything around it.

Visual Cliché

Despite *Gravity*'s mesmerising cinematography the film remains a blockbuster space adventure. When watched in its original 3D format, *Gravity* includes a great deal of stomach-churning, vertiginous action. Cultural historian Kodwo Eshun wrote of his experience of parabolic flight as 'metaphysics coming down to earth and into the mouth' (Eshun, 2005, p. 28); while *Gravity* has slow, contemplative moments, for the majority of the film we experience what could be described as cinema in the mouth. It is a visceral ninety minutes. However, Cuarón's aspiration for the film went beyond creating an action-packed survival drama. He states 'it's a metaphorical tale above anything else,' a tale told 'not in a rhetoric way but through visual metaphors' (*Behind the Scenes*, 2014).

To tell the metaphorical tale, one of 'rebirth as an outcome of adversities' (Cuarón in *Behind the Scenes*, 2014), the film draws upon a number of visual motifs. Images of astronaut/foetus and mother/Mother Earth seem to point to this theme of 'rebirth'. In the case of the protagonist, Stone, this is a metaphorical 'journey back into life' (Cuarón in *Behind the Scenes*, 2014) after a period of mourning and emotional ossification caused by the death of her child. The echoing motif of Stone's face overlapped with images of Mother Earth seem intended to align her physical survival with the character's emotional resurgence; one entry in the shooting script reads: 'The Earth glimmers brightly. A third of the hemisphere is in complete darkness. Her FACE REFLECTED in the glass is SUPERIMPOSED over the EARTH' (*Behind the Scenes*, 2014).

However beautifully realised, this and the other visual motifs Cuarón deploys never go beyond the surface aesthetic, resulting in laboured visual cliché as opposed to deep metaphorical resonance. Richard Brody writing in *The New Yorker* states: 'It's hard to recall a movie that's as viscerally thrilling and as deadly boring as *Gravity*' (Brody, 2013). Perhaps, as Brody seems to imply, the flaw in Cuarón's film is that the metaphorical tale stands at odds with high-speed action that keeps the viewer in a state of high alert throughout.

There is one moment in the film in which the transformative possibility of *Gravity*'s spellbinding spacescape does materialise convincingly. This is conveyed through Kowalski and not Stone. It occurs in the opening sequence of the film, an astounding tracking shot of around 13 minutes. As I have touched upon this is the shot in which Kowalski marvels at the view from 600 km 'up'. First we see the astronaut's head to the left of the screen with a blue, Earth-shaped reflection on his visor. Looking up through the reflection he watches the Earth itself gently float down into shot. The camera then glides away from Kowalski's face tracking slowly across the skin of the Earth. To conclude this beautiful 360-degree fluid pan Kowalski's face appears once more, but this time to the right of the screen. Struggling to find words to describe what he sees, a sense of awe apparent in his eyes, he simply murmurs the single word 'terrific'. The innovative way in which the film is lit allows the subtlest changes of light within the digital scenography to play out across the actor's visor, skin and in the liquid luminosity of the actor's eyes, but there is also something more than technical brilliance in this one emotive moment.

The technical innovation of Cuarón's film resulting in both a sense of thrill and boredom (Brody, 2013) reminds me of Werner Herzog's subtle critique of NASA in his science fiction fantasy, *The Wild Blue Yonder* (2005). In a section entitled 'Utopia of the Ideal Colony', Herzog interviews Martin Lo, a research scientist working in NASA's jet-propulsion laboratory. Lo appears to respond to a question about what might happen once mankind finds a way to travel to another habitable planet. The scientist talks of building a space colony, a large dome in an Amazonian setting. He suggests that the main feature might be 'a shopping mall in space, because you have everything, you could shop all day long, this might be... the perfect space colonisation paradigm' (Herzog, 2005). As Lo finishes speaking, Herzog cuts back to a deserted town and the actor Brad Dourif, playing an alien who landed on Earth many moons ago, standing in front of his own failed shopping mall. Dourif whispers in despair, 'shopping mall! I could have told them... it makes me so sad' (Herzog, 2005). Herzog's film is about the incredible technical imagination required within the space programme and, at the same time, the dearth of cultural imagination when it comes to the radical possibilities it opens up. In the making of *Gravity*, Cuarón has achieved something technically brilliant but the metaphorical tale that unfolds is almost as mundane as a shopping mall.

Radical Landscape

The flaws within Cuarón's film should not distract from the fact that *Gravity*'s beautiful spacescape is unlike anything we have seen before. The film's cinematography, created by painstakingly unpicking and unlearning the instinct to show things as they appear according to deeply embedded convention of image-making, seems to dismantle the genre of landscape itself. Just as the tiny fragments of debris moving at high speed in near-Earth orbit cumulatively disabled the space shuttle Explorer within the narrative of the film, frame by frame *Gravity*'s cinematography dismantles the constraints of terra firma image-making. How can a radical world in which everything moves in all directions not make us consider the limitations of landscape as we understand it today?

In his introduction to a collection of essays *Landscape and Film* (2006), Martin Lefebvre appears to recognise the inadequacies of landscape as a genre by suggesting that it has become a kind of catch-all. He writes, 'landscape remains notoriously difficult to define, having apparently no single set of fixed criteria outside of its spatial nature' (Lefebvre, 2006, p. xiii). He argues that despite this, one thing is very clear: 'in investigating landscape in film one is considering an object that amounts to much more than the mere spatial background that necessarily accompanies the depiction of actions and events' (Lefebvre, 2006, p. xii).

While specifically about film, Lefebvre's argument can be applied to landscape in general regardless of whether it is filmed, photographed, drawn or painted. Even the 'real' landscapes documented within NASA's (2019) extensive archives need be understood as no neutral backdrop to what could be described as the actions that constitute history, but rather considered as images that (have been used to) shape history.

Upon first viewing *Gravity* it is evident that Cuarón's radical planetary landscape is no neutral backdrop to the action of the film. As I have drawn out, it is intended to actively shape the narrative, specifically *Gravity*'s Earth, which Cuarón describes as 'an ever-present character' moving around within the cinematic frame. What makes *Gravity*'s Earth even more noteworthy is the role it plays in a different narrative, that of mankind's encounter with images of Earth, viewed from outside, as traced within mainstream culture today.

The Second Space Age

Images of Earth have mesmerised mankind since the late 1960s. In his book *Earthrise: How Man First Saw the Earth (2008)*, Robert Poole tells this story. He describes the period just after the Apollo years as one in which the image of the whole Earth was everywhere; 'it seemed to some to mark "a new phase of civilisation", the beginning of the "age of ecology"' (Poole, 2008, p. 9). This was the era in which photographs such as *The Blue Marble* (1972) led to the awakening of our 'planetary consciousness' (Sagan, 1994, p. 171). However, as William E. Burrows explains, we are now in a new era, in what he identifies as the second space age. Burrows draws out the fact that the first space age is defined by 'daring and soul-stirring feats of exploration to both the Moon and the planets', i.e. *Apollo* (1968–1972), *Viking* (1975–1983) and *Voyager* (1977–). These are journeys away from the Earth and in the case of Voyager 2, this journey continues as it travels on through interstellar space.

In comparison, the second space age is defined by 'the use of space, and particularly low Earth Orbit, to serve society's needs' (Burrows, 1999, p. 610). Burrows describes the key trait of this second phase as the institutionalisation of the space programme into our daily lives enabled by the 'constellation of satellites' that are 'in orbit around the Earth' (Burrows, 1999, p. 617). The space shuttle programme (1981–2011), the Hubble telescope (1997–) and the International Space Station (1998–) are the key markers of shift from the first to the second space age. In the late 1990s the latter two took up residency in near-Earth orbit where they have been feeding us with images of deep space and planet Earth, respectively. British Astronaut Tim Peake, who spent six months on the International Space Station, writes: 'It's impossible to look down on Earth from space and not be mesmerised by the fragile beauty of our planet ... I became determined to share this unique perspective of the one place we all call "home"' (Peake, 2016, p. 10).

The Earth that Tim Peake photographed is the Earth featured in Cuarón's *Gravity,* for the setting of *Gravity* is the near-Earth orbit of the second space age. As I have drawn out, the view from this orbit shows a planet that is partial, beguiling and often ethereal in appearance due to the blue halo that separates the curved horizon from the black abyss beyond. In comparison, the Earth of the first space age, the images that saturated the visual culture of the twentieth century, are of the whole Earth. *The Blue Marble* photograph features the Earth shot from a distance of around 29,000 km. This distance is too great to see the local scattering of blue, hazed atmosphere that give the near-Earth shots their ethereal appearance. In comparison the whole Earth is clean-edged. Its aesthetic attributes are those of a perfect, marbled sphere.

Robert Poole draws out the fact that the Earth of the Apollo years became the ultimate scientific object: 'Like a space-age version of Newton's windfall apple, the image of the whole Earth fell ripe from orbit in full view of a scientific public ready to receive it' (Poole, 2014, p. 216). Arguably, these images marked a moment at which Earth could study itself, could gather itself up in what Michel de Certeau might identify as the totalising imaginary.

Not only did the whole Earth transfix scientific imagination at this time, but its clean-edged aesthetics seemed perfectly in keeping with the progressive ideals of the modern era in general. It was at one with the formalism of a Mies van der Rohe building, a Barbara Hepworth sculpture and, of course, a Stanley Kubrick film. As it appears in both the opening and closing sequence of *2001,* Kubrick's Earth has the precise curvature of a modernist sculpture. Working prior to Apollo 8's 'first recognisable views of the whole Earth' (Poole, 2008, p. 21) *2001*'s special effects team painted a planetary image direct on to glass, and then 'back-lit Earth was photographed on an animation stand' (Agel, 1970, p. 89). 'Transparencies needed careful exposure to make Earth look like (a) bright object' (Agel, 1970, p. 98). The resultant cinematic portrait perfectly anticipated the glowing, modernist marble photographed by William Anders (1968), Eugene Cernan (1969), and Ronald Evans and Harrison Schmitt (1972).

This vision of our blue, marbled planet, as imagined in *2001* and documented by the Apollo Mission, informed the way we represented our planetary home for decades to come. It is not until the late 1990s that we start to see change. At this time the second space age, with its shifted viewpoint, begins to infiltrate the stories we tell ourselves about space and the images that go with these stories. Within the genre of film we can trace MIR Space Station featuring in a cameo role in *Contact* (Robert Zemeckis, 1997) and *Armageddon* (Michael Bay, 1998); and The International Space Station appearing in *The Day After Tomorrow* (Roland Emmerich, 2004) and *Love* (William Eubank, 2011). The brief shots of Earth that we see in these films are of a partial, blue-limbed planet.

Furthermore, in 2012 Universal Pictures celebrated its centenary by changing its logo from a clean-edged globe to something more contemporary, a near-Earth shot of a blue-hazed celestial form. The following year Universal released Joseph Kosinski's *Oblivion* (2013), Tri-Star released Neil Blomkamp's *Elysium* (2013) and Warner Bros. released Alfonso Cuarón's *Gravity* (2013). These comprise a trilogy of near-Earth science fiction films all featuring a beautiful, blue-limbed planet, with *Gravity*'s Earth being the most notable of the three. The epic nature of *Gravity*'s portrait of Earth, the length of time it swims before us on the screen demanding our attention with its luminosity and agility, forces us to take stock of this planetary portrait and consider it in relation to those views from Apollo 8, 'you are looking at yourself from 180,000 miles out in space'. It makes us look around and see that the images of Earth that saturate contemporary visual culture are no longer the modernist marble of the first space age but show Earth as near, partial and blue-hazed.

Evidently this change is not absolute. In the first space age we saw shots of Ed White floating above an azured, partial Earth. Today, despite the dearth of new images of the whole Earth, *The Blue Marble* and composite satellite images are frequently reproduced. However, the shifting viewpoint of the second space age has given us a 'new' default image of Earth, which has saturated mainstream culture. It has given us a new cultural imaginary of Earth represented by way of a new set of aesthetics. *Gravity*'s cinematic portrait firmly draws our attention to this change.

The New Earth

This new planetary presence that swims before us no longer appears as if lost in the immensity of a voided sky, a lonely jewel wrapped in darkness, but its enormous, celestial form is immanent as it floats in a halo of blue. This new Earth no longer makes meaning as the perfect scientific object, one which satiates the desire of a totalising imaginary. Instead, its elusive whale-like body awakens our senses to a different kind of pleasure, a different kind of imagination.

Within contemporary philosophy there are voices that call out for a renewal of ideas and imagination. Franco 'Bifo' Berardi is one such voice. He speaks of the need for a new language, and arguably new images (2017). Just as the whole Earth did before, surely this new image of Earth, this depiction of a looming spectral presence swimming through our 21st-century skies, also has the power to unleash new ideas and alter the course of history?

Bibliography

Agel, J. (ed.) (1970), *The Making of Kubrick's 2001*. New York: New Amer Library.

Berardi, F. (2017), 'The Great Experiment – Franco "Bifo" Berardi in conversation with David Campany and Laura Mulvey'. Central Saint Martins, 5 October [Public Lecture].

Bizony, P. (1994), *2001: Filming the Future*. London: Aurum Press Limited.

Brody, R. (2013), 'Generic Gravity', *The New Yorker*, 4 October [online],https://www.newyorker.com/culture/richard-brody/generic-gravity, (accessed 10 February 2019).

Burrows, W.E. (1999), *This New Ocean: The Story of the First Space Age*. New York: The Modern Library.

de Certeau, M. (1984), *The Practice of Everyday Life*. Berkeley, Los Angeles & London: University of California Press.

Eshun, K. (2005), 'On the Use and Abuse of Microgravity for Life' in N.Triscott & R. La Frenais (eds), *Zero Gravity: A Cultural User's Guide*. London: The Arts Catalyst, pp. 28–32.

Kessler, E. (2012), *Picturing the Cosmos: Hubble Space Telescope Images and the Astronomical Sublime*. Minneapolis: University of Minnesota Press.

Lefebvre, M. (2006), 'Introduction' in M. Lefebvre (ed.), *Landscape and Film*. New York & London: Routledge, pp. xi-xxxi.

NASA (2019) [online], www.nasa.gov (accessed 10 February 2019).

Nietzsche, F. (1974), *The Gay Science*. New York: Vintage Books.

Peake, T. (2016), *Hello, is this planet Earth? My View From the International Space Station*. London: Penguin Random House.

Pond, S. (2013), 'Lost in Space with Sandra Bullock: Alfonso Cuarón's 10 laws of "Gravity"', *The Wrap* [online], www.thewrap.com/lost-in-space-alfonso-cuarons-10-laws-of-gravity/ (accessed 14 April 2019).

Poole, R.(2014), 'What Was Whole about the Whole Earth? Cold War and Scientific Revolution' in S. Turchetti, and P. Roberts, ed. *The Surveillance Imperative: Geosciences during the Cold War and Beyond*. London: Palgrave Macmillan, pp. 213–235.

Poole, R.(2008), *Earthrise: How Man First Saw the Earth*. New Haven & London: Yale University Press.

Radford, I. (2014), 'A trip through Gravity's extras: iTunes vs Blu-ray', *VODzilla.co* [online], https://vodzilla.co/blog/features/a-trip-through-gravitys-extras-itunes-vs-blu-ray/ (accessed 10 February 2019).

Sagan, C. (1994), *Pale Blue Dot: A Vision of the Human Future in Space*. New York: Random House.

Virilio, P. (1997), *Open Sky*. London: Verso.

Woods, D.W. & O'Brien, F (eds) (2021), Day 4: Lunar Orbit 4, *Apollo Flight Journal*, [online] https://history.nasa.gov/afj/ap08fj/16day4_orbit4.html (accessed 10 February 2019).

Filmography

2001: A Space Odyssey (1968) [Film], Dir: Stanley Kubrick, UK & USA: Stanley Kubrick Productions.

Apollo 13 (1995) [Film], Dir: Ron Howard, USA: Universal Pictures & Imagine Entertainment.

Elysium (2013) [Film], Dir: Neill Blomkamp, USA: TriStar Pictures, Media Rights Capital, QED International, Alphacore & Kinberg Genre.

Gravity (2013) [Film], Alfonso Cuarón, UK & USA: Heyday Films, Esperanto Filmoj.

Gravity Special Feature: Behind the Scenes (2014) [DVD], USA: Warner Home Video.

Gravity Special Feature: Collision Point: The Race to Clean Up Space (2013) [DVD], Dir: Brian Macdonald, USA: Warner Home Video.

Oblivion (2013) [Film], Dir: Joseph Kosinski, USA: Universal Pictures.

Sunshine (2007) [Film], Dir: Danny Boyle, UK & USA: Moving Picture Company, DNA Films, UK Film Council & Ingenious Film Partners.

Star Wars (1977) [Film], Dir: George Lucas, USA: LucasFilm Ltd.

The Wild Blue Yonder (2005) [Film], Dir: Werner Herzog, UK, France & Germany: Werner Herzog Filmproduktion, West Park Pictures Produktion, & Tétra Média.

Chapter 2

A *Short Vision*'s reception and perception: the story of a film that "came unnoticed, uninvited"

Sara Khalili

I n this chapter, I conduct a contextual study of *A Short Vision*, an animated short by Peter and Joan Foldes produced in 1956 in Britain. In order to do so, I draw from Klinger (1997) on factors related to reception studies as a framework to examine briefly how the film was received upon release as well as in recent decades. I will explore how *A Short Vision* morphs into multiple visions, and how an implicit layer or theme of 'atomic war' explicates boldly on contemporaneous events. I will discuss film distribution, exhibition, cross-cultural reception, plus biased writings and reports about the film.

A Short Vision, according to different claims, synthesized many of the features of the early Cold War and created a different vision of nuclear war. It was named as one of the most influential animations capturing the mood of nuclear paranoia. However, it appears that the 'atomic war', as a main theme, was injected into discourse on the film just after it screened to shocking effect on the Ed Sullivan Show in 1956, and Sullivan's particular way of introducing *A Short Vision* to the audience. This case study is a tangible example of how the context might be crucial in shaping the meaning and/or the identity of a film, and how even a short experimental non-mainstream animation could reflect on the realities about time, mood, and other conditions in which it has been created, received, and perceived.

Introduction

The purpose of contextual studies is to understand how the contextual components of a film might reveal, create, or manipulate the meaning of it. According to Klinger (1997), reception studies examine the nested relationships of a film and its contextual areas such as distribution, exhibition practices, reviews, and the dominant or alternative ideologies of the society at a particular time. Vasudevan (2010) confirms history is distributed into various elements implying several histories; this means that an under-standing of a film in history emerges from its dispersal. Klinger (1997) identifies two categories of synchronic and diachronic areas of study to create a space to better understand the context of a work. Synchronic areas include cinematic practices, intertextual zones, and social-historical contexts; while diachronic areas relate to revivals, reviews, biographical legend, etc.

I will focus on a few areas among synchronic elements as well as diachronic areas to deepen understanding of the film context. To do so, it is necessary briefly to explain the narrative and style of the film as an introduction to its contextual exploration.

Narrative and Style

The film starts with a line transforming into a missile-like shape in the night sky, passing over the sleeping city, causing the scared animals to pause and stop fighting with each other. The 'leaders' and the 'wise men' look upwards but it is 'too late'. Their knowledge seems useless and the whole world is destroyed by the exploding object. The only remaining things at the end are a small flickering flame and a fluttering moth circling it, both of which will die soon.

According to Jacobi, the film 'synthesized many of the features of the early Cold War aesthetic' (Jacobi, 2014, p. 41). Brooke (2003) interprets the narration as an absolute extinguishing of any hope at the end. He mentions its quasi-Biblical mood, talking about a mysterious 'it' that appears in the sky terrifying animals but is ignored by humans. Grant (2013) claims the utilization of a noble narrative technique to create a different vision of nuclear war, which rejects survival and includes death and destruction. Similarly, Pitetti confirms that this dark 'anti-nuclear film' represents a new eschatological narrative different from older religious apocalyptic models (Pitetti, 2016, p. 62).

Regarding the visual style, Jacobi associates the absolutely tough image of nature, the dark city, the exaggerated faces, and the dying flame and dark-ness at the end – all performed in a 'sudden', 'basic' graphic fashion – with abstract expressionist style (Jacobi, 2014, p. 41). At the same time, Beylie (1964) argues for the influence of Picasso and Rouault in *A Short Vision*.

Synchronic Study

Synchronic areas of study, according to Klinger (1997), are described as:

> [M]ost closely associated with the production of a film [...] moving to those technically outside the industry, but closely affiliated with a film's appearance [...] and ending with social and historical contexts circulating through and around its borders (Klinger, 1997, p. 113).

Relying on synchronic areas, I articulate related factors including film

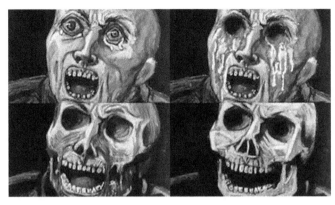

Figure 1: *A Short Vision* (1956), Peter and Joan Foldes, Britain [online], available at: https://marcianosmx.com /broadcast-interruption-fin-la-humanidad/ (accessed 14 December 2017).

personnel, film production and distribution, journalism, social and histori-cal contexts, and law and ethical considerations.

Film Personnel, Production, and Distribution

As Klinger (1997) mentions, the studio's choice of crew has a lively impact on a film's social recognition and appropriation. He adds that providing a production history for the film would help to grasp how its characteristics were getting shaped by the economic structure and production practices of the studio (Klinger, 1997).

In the case of a short experimental animation, the director is usually considered the most important of the personnel. *A Short Vision* was Peter Foldes' and his wife Joan's second and last animation together. Peter Foldes (1924–1977) was a Hungarian–British artist who started with *Animated Genesis* (1952), a well-recognised work in the field of experimental anima-tion.[1] *A Short Vision* was narrated by the British actor, James McKechnie, and was accompanied by haunting futuristic music by Màityaìs Seiber, who was a well-known Hungarian composer. The film won the *Prix pour la couleur* at the Cannes Film Festival and a special award at the British Film Academy awards.

A Short Vision was produced in the Foldes' kitchen (Manvell, 1961) in Edgware. It was made as a private venture in the beginning but was com-pleted with the help of the British Film Institute (BFI)'s Experimental Production Fund, which sponsored approximately thirty animations be-tween the mid-1950s and 1990s (Brooke, 2003). *A Short Vision* was distrib-uted through British Lion Films (Huntley, 1956).

Focusing on the arena of film distribution highlights the way the film might have been distributed and the conditions under which it might have ap-peared theatrically (Klinger, 1997). In the case of *A Short Vision*, there was a critical financial situation in the funding system at the time during which the film was produced. However, it had perhaps the consequence of encour-aging the BFI to generate revenue from the distribution and sales of the

Figure 2: *EdScreen& AV Guide,* June 1957, US [online], available at: http://archive.org/stream/ educationalscree36chicrich#page/n319/mode/2up (accessed: 14 December 2017).

films to a US television company for primetime broadcasting (Dupin, 2003). The generated revenue over the next ten years reached over £10,000, which was income during the Experimental Film Fund's dying years (1963–1966).

Geerhart clarifies that before the film had its historical screening in the US, it was shown at the National Film Theatre in London and reviewed in the London *Times* on January of 1956 under the headline 'Cartoon of the End of the World' (Geerhart, 2011). The advertisement above, which appears on June of 1957 in an American magazine, is referring to this review.

Regarding the film's journey from England to *The Ed Sullivan Show* in the US, Beck (2011) points out that '[t]he fact that a mainstream U.S. variety show ran this art-film-with-a-message in primetime is almost as shocking as the film itself' (Beck, 2011). Sullivan's show used to host comedians, acrobats, and rock stars like Elvis Presley, the Rolling Stones, and the Beatles. However, and over the years, it tended to host intellectuals or artists like Carl Sandburg, Salvador Dali, and Maria Callas as well. Ed Sullivan, who after cancelling his show by CBS in 1971 was recognised as a gossip columnist in the New York *Daily News*, claimed he showed *A Short Vision* intentionally to help world peace, while Geerhart (2011) suspects he may have decided to run the film for business reasons, since he had a relationship with its US distributor, George K. Arthur, and announced in advance its commercial distribution in the fall (in the US) by George K. Arthur. Geerhart also mentions, 'the host's evolving Cold War attitudes may have

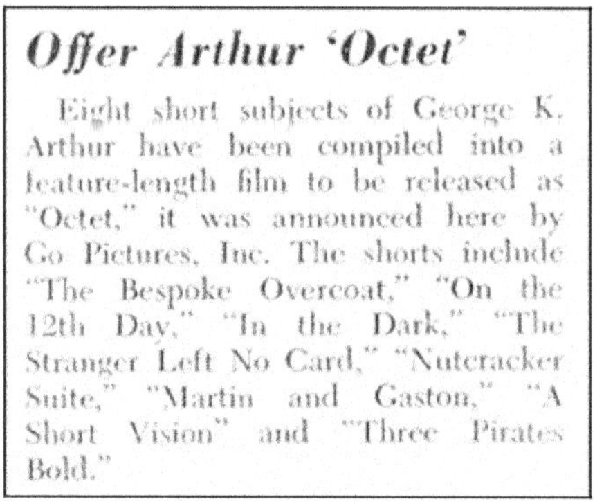

Offer Arthur 'Octet'

Eight short subjects of George K. Arthur have been compiled into a feature-length film to be released as "Octet," it was announced here by Go Pictures, Inc. The shorts include "The Bespoke Overcoat," "On the 12th Day," "In the Dark," "The Stranger Left No Card," "Nutcracker Suite," "Martin and Gaston," "A Short Vision" and "Three Pirates Bold."

Figure 3: *Motion Picture Daily,* March 28, 1956 [online], available at: https://lantern.mediahist.org/ catalog/motionpicturedai83unse_0392 (accessed 14 December 2017).

also played a part' in airing this groundbreaking animation that was labeled an 'anti-war cartoon' (Geerhart, 2011) Regardless of the 'true' motivations for running this film in the US, Sullivan's program still seems an odd choice in which to broadcast this animation.

From another perspective (Gault, 2016), Sullivan was a trusted media figure who abused that trust by scaring a generation of children about/of the atomic war, particularly by remembering that the film to some extent reflects a sense of a child's storybook or biblical lesson through its

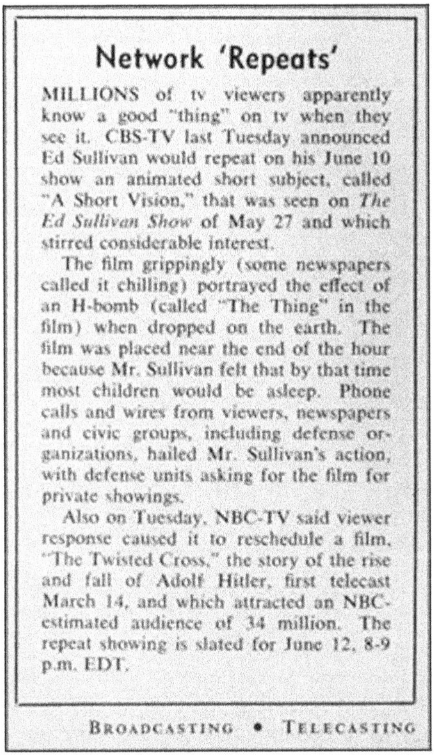

Network 'Repeats'

MILLIONS of tv viewers apparently know a good "thing" on tv when they see it. CBS-TV last Tuesday announced Ed Sullivan would repeat on his June 10 show an animated short subject, called "A Short Vision," that was seen on *The Ed Sullivan Show* of May 27 and which stirred considerable interest.

The film grippingly (some newspapers called it chilling) portrayed the effect of an H-bomb (called "The Thing" in the film) when dropped on the earth. The film was placed near the end of the hour because Mr. Sullivan felt that by that time most children would be asleep. Phone calls and wires from viewers, newspapers and civic groups, including defense organizations, hailed Mr. Sullivan's action, with defense units asking for the film for private showings.

Also on Tuesday, NBC-TV said viewer response caused it to reschedule a film, "The Twisted Cross," the story of the rise and fall of Adolf Hitler, first telecast March 14, and which attracted an NBC-estimated audience of 34 million. The repeat showing is slated for June 12, 8-9 p.m. EDT.

BROADCASTING • TELECASTING

Figure 4. *Broadcasting,* 4 June 1956 [online], available at: https://lantern.mediahist.org/catalog/broadcastingt ele50unse_0_1264 (accessed 9 June 2020).

narration and visual style. Rerunning the film two weeks later after all its shocking influence is interpreted by Gault as an intention to horrify those who missed it the first time; Sullivan and CBS heavily promoted the second airing of the film by advertising in newspapers and even inviting the Navy Blue Angels to the studio (Gault, 2016).

The film toured in festivals and won awards (such as the best experimental film at the 17th Venice Film Festival in September 1956) after airing on *The Ed Sullivan Show*.

Journalism

According to Allen and Gomery (1985), film criticism in newspapers, magazines, radio and television, define the terms by which a film will be discussed and evaluated in public (Allen & Gomery, cited in Klinger, 1997). In the table below, I summarized coverage of *A Short Vision* in a few newspapers and magazines in the UK and the US, from January to July of 1956. As briefly mentioned, *A Short Vision* was aired twice in Sullivan's primetime show: first on 27 May 1956, and then on 10 June 1956. As Geerhart (2011) states, '[u]nlike its stealth airing on the May 27 [...] *A SHORT VISION* was heavily promoted in newspaper programming highlights and television listings for its June 10th reprise'.

Crafton reminds us to distinguish between a film's social context and the writings of those with biases or interests in laying claim to the film for their own purposes (Crafton, 1996). Therefore we can explore how *A Short Vision* morphs into multiple visions (see Table 1) and how an implicit theme of 'atomic war' unfolds, particularly after its appearance on *The Ed Sullivan Show*. Two examples of the newspapers can be found in Figures 5 & 6.

It is worth highlighting that during *A Short Vision*'s initial release in the UK, the film was described in terms unrelated to atomic warfare. After *Ed Sullivan* identified the film as about atomic war, it started being recognised and sparked debates within that specific context. Within six months of its initial premiere in the UK, the film had become firmly associated with nuclear affairs in US magazines and in July 1956 was labeled similarly in the

AND SUN, MONDAY, MAY 28, 1956

Shock Wave From A-Bomb Film Rocks Nation's TV Audience

Toast of Town Offering Draws Criticism, Praise

By CAROL TAYLOR, Staff Writer.

To some it was "seven minutes of terror." To others it was "the best piece of anti-war propaganda ever shown."

Such was the reaction of millions of viewers last night to a chilling cartoon film depicting the destruction of the world by atomic warfare. But almost all

A woman's face as affected by the Thing, an atomic bomb that destroys the world. The drawings show, left to right, before, during, and after the Thing strikes.

Figure 5. *NY World-Telegram & Sun,* 28 May 1956 [online], available at: https://flashbak.com/a-short-vision-the-1950s-nuclear-armageddon-cartoon-that-terrified-everyone-watching-the-ed-sullivan-show-36372/ (accessed 9 June 2020).

UK. Remarkably, between August 1956 and July 1957 a gradual shift in its portrayal could be observed in US journals and newspapers.

Social and Historical Contexts

The mid-1950s was a period of international nuclear terror. Both America and the Soviet Union had increasingly stronger warheads. In 1952, the first hydrogen bomb was successfully detonated and a new wave of fear swept through the world. Jacobi (2014) refers to the dramatisation of the global inevitability of radiation spread in Nevil Shute's novel, *On the Beach*, prefaced by T.S. Eliot's words 'This is the way the world ends' (Jacobi, 2014, p. 40). Similarly, American films such as *Invasion of the Body Snatchers* (Don Siegel, 1956) and *The Incredible Shrinking Man* (Jack Arnold, 1957) reflected the mood of the time, nuclear paranoia (Brooke, 2003). According to Brooke (2003), there is though no explicit reference to atomic warfare within *A Short Vision.* As explained earlier, this as a 'main' theme is present in reception of the film after Sullivan's show and his introduc-

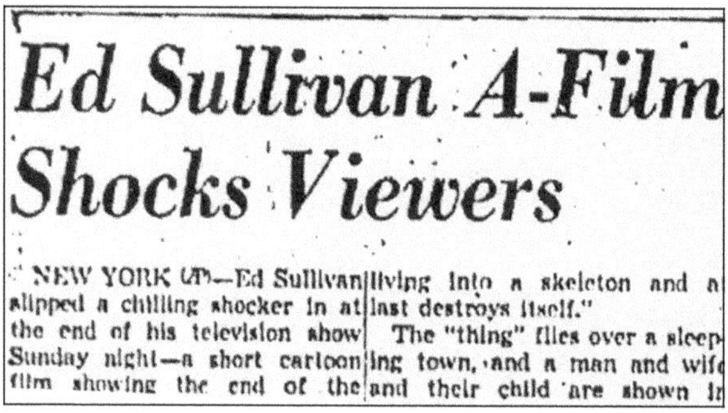
Ed Sullivan A-Film Shocks Viewers

NEW YORK (AP)—Ed Sullivan slipped a chilling shocker in at the end of his television show Sunday night—a short cartoon film showing the end of the living into a skeleton and at last destroys itself."

The "thing" flies over a sleeping town, and a man and wife and their child are shown in

Figure 6. *Associated Press,* 29 May 1956 [online], available at: http://conelrad.blogspot.com/2011/06/short-vision-ed-sullivans - atomic-show.html(accessed 9 June 2020).

Date	Publisher	Country	Highlights
A Short Vision in newspapers and magazines (US & UK)			
Jan 26 1956	The London Times	UK	"Cartoon of the End of the World" A work of "sombre imagination"
May 28 1956	NY World-Telegram & Sun	US	Sullivan's "plea for peace"
May 29 1956	Associated Press	US	"Ed Sullivan A-Film Shocks Viewers" "End of the world by atomic warfare" "heavy reaction" (pro and con)
May 30 1956	Post-Standard (Syracuse, NY)	US	"Anti-War Document" (editorial) Bringing home the reality of atomic warfare Sullivan plans to re-air it on June 10 Should be shown in Russia and everywhere Best piece of anti-war propaganda ever shown Shocked public reaction was natural
May 31 1956	Los Angeles Times	US	Sullivan got an enormous amount of mail to repeat it
June 11 1956	Time magazine	US	So chilling even in black and white The studio audience stunned when it was over Phone calls divided between praise & condemnation
June 11 1956	NY World-Telegram & Sun	US	"Shock Wave From A-Bomb Film Rocks Nation's TV Audience." Received rave notices from London Times and Manchester Guardian and Sullivan gave it a US premiere
June 12 1956	Hollywood Reporter	US	Popular demand to repeat the show Drawings of H-Bomb effect especially realistic
July 1956	The Monthly Film Bulletin, The British Film Institute	*UK*	Imaginative and disturbing picture of atomic warfare The horror presented unflinchingly on American TV Terrible close-ups of decomposing faces
Aug 1956- Jul 1957	North Adams, New York Times, Eagle, The Saturday Review, etc.	US	More modest reviews: "A beautiful and bloodcurdling little animated picture" "Admirable but not for the meek"

Table 1. Coverage of *A Short Vision* in newspapers and magazines (US & UK), January to July of 1956.

about nuclear conflict. This mainstream reading of the film begins just after his claim that everybody takes it for granted that the film is about atomic warfare.

Law and Ethical Considerations

As Klinger (1997) states, the aim here is to place a film 'within the framework of legal rulings' (Klinger, 1997, p. 119). According to Geerhart (2011), even a black and white broadcast of *A Short Vision* could have traumatized a generation of children and would probably freak out today's more sophisticated five-year-olds. Sullivan's parental advisory before airing *A Short Vision* for the first time might have even attracted the curi-

Figure 7. Ed Sullivan superimposed by the end title of *A Short Vision* [online], available at: http://conelrad.blogspot.com/2011/06/short-vision-ed-sullivans-atomic-show.html (accessed 9 June 2020).

osity of the younger audience: 'I'm gonna tell you if you have youngsters in the living room tell them not to be alarmed at this 'cause it's a fantasy, the whole thing is animated.' After the film was done, Sullivan looked 'knowingly' at his audience, smiled and said 'See', which might have meant, 'I told you so' (Geerhart, 2011).

Sullivan might have underestimated the traumatic effect the film would have on children but after the show he received huge criticism for this. Regarding ethical considerations and regulation of television, it is perhaps surprising that such material had been broadcast so easily, and this might have caused him to modify his wording when airing the film for the second time on 10 June:

> It is a harrowing experience for youngsters, so would you please take them out of the room and just have the older people in the family look at it. I think it's something the country should know, should see, but the youngsters, that is the little ones, should not be looking at it (cited in Geerhart, 2011).

Diachronic Study

According to Klinger (1997), '[d]iachronic research [...] forces consideration of a film's fluid, changeable and volatile relation to history' (Klinger, 1997, p. 112). A few factors from her suggestive mapping significant to diachronic study are selected and explained as follows.

Revivals

The appearance of a film from the past in revival houses, art museums, festivals and archives, according to reception theories, can generate new meanings (Klinger, 1997). Since 1956, *A Short Vision* has been mentioned every now and then in books and articles about animation and film, and more notably, recently on websites and blogs, particularly a recollection of it by CONELRAD, which I will discuss below.

Reproduction

A 16mm print of the decidedly anti-war film is stored in the Federal Emergency Management Agency records (Geerhart, 2011). This print is an American distribution copy with the credit 'George K. Arthur Presents' and is not of the best quality. CONELRAD (Control of Electromagnetic Radiation) in the US obtained a colour copy of the film from the National Archives in College Park, Maryland. The copy of *A Short Vision* (in colour) that was posted on YouTube in May of 2009 by the BFI is the first online copy of the film and prior to this, the BFI restricted watching it to people in the UK only with library accounts. Geerhart reports that he was not able to find any entry about the film on IMDB or in Wikipedia in May 2009 (Geerhart, 2011). It is worth mentioning that when writing the first draft of this paper on December 2017 and doing a rough search on the internet, I came to more than 30 blog entries and a few reviews in IMDB, which could be a result of reproduction of the film in 2009 and particularly highlighted by CONEL-RAD in June 2011.

Fan Culture

Fan culture extends beyond the original moments of reception, lasting long afterwards to create group identities around media products, as well as develop discussions. This is then a powerful area for considering how the media may be reactivated over time by specific groups with particular social identities and interpretive agendas (Taylor, 1989).

In this context, CONELRAD – which describes itself as a website focused on 'atomic secrets, missing persons, and general Cold War strangeness' – developed a project called 'A SHORT VISION: LEGACY PROJECT' in 2011 and published an extensive history of the film in American broadcast and the subsequent media reaction. It invited enduring childhood memories of the baby boomers who remembered *A Short Vision* from *The Ed Sullivan Show*. In this project, stories of the 'Fifties kids' who remembered seeing the 'world end' on TV back in 1956 were gathered.[2] Most of these people indicate that watching this film in their childhood strongly influenced them; there are people who talked about their nightmares afterward. This collection is a potentially rich resource for oral history. However, analysing these narratives is beyond the scope of this paper.

The Biographical Legend

As Klinger states, 'The biographical legend of film personnel begins in the synchronic moment, but through the cumulative effects of time achieves potentially its biggest impact on meaning' (Klinger, 1997, p. 126). Peter Foldes' pioneering computer animation, *Hunger* (1974), which explored 'free distortion and metamorphosis with the new automated movement' (Tai, 2013, p. 111), was nominated for the Best Animated Short Film at the 47th Academy Awards. His other work, *Rêve* (1977), won the 1978 César Award in France after his death. The 1977 Cannes Film Festival, which took place several months after Peter Foldes passed away, held a special homage for his body of work in animation. These realities which make him a legend may affect one's interpretation and perception of his past works including *A Short Vision*.

Conclusion

A Short Vision, as one of the most powerful animations created in the UK during the Cold War, as a consequence of some surrounding events (synchronic and diachronic), transformed into a recognised animation with the main theme of 'nuclear war'. In this chapter, I aimed to show how the context might shape a new meaning or identity for a film, and foreground a shadow-sighting theme in the stratosphere.[3] At the same time, this case study shows how even a short experimental animation can uncover realities about the time, mood, and other conditions in which it has been created, received and perceived. In this paper, I only had the chance to scratch the surface and briefly explore a few elements based on Klinger's framework in reception studies; each section could potentially be focused further and dug deeper in order to create a space to understand the film and its context better.

Endnotes

1. This animation shares some thematic and artistic touches with *A Short Vision,* for example, a big spider (evil) chasing a moth (good) (Stephenson, 1973).

2. http://conelrad.blogspot.ca/2011/06/short-vision-legacy-project.html

3. I wanted to refer to the symposium title, Beasts of the Sky: Strange Sightings in the Stratosphere, in which a part of this paper was presented.

Bibliography

Allen, R.C & Gomery, D. (1985), *Film History: Theory and Practice,* New York: Knopf.

A Short Vision (1956) [Film], Wr/Dir: Peter and Joan Foldes, UK, 6:09 minutes.

Beck, J. (2011), 'A Short Vision', *Cartoonbrew* [online], http://www.cartoonbrew.com/shorts/a-short-vision-by-joan-and-peter-foldes-45369.html, accessed December 2017.

Beylie, C. (1964), 'La peuratomique à l'écran',*Séquences: La revue de cinéma*, Vol. 39, pp. 4–13.

Brooke, M. (2003), '*Short Vision, A* (1956)', *BFI Screenonline*, http://www.screenonline.org.uk/film/id/442000/, accessed December 2017.

Crafton, D. (1996), '*The Jazz Singer*'s Reception in the Media and at the Box Office' in D. Bordwell & N. Carroll (eds), *Post-Theory: Reconstructing Film Studies*, Madison, WI & England, University of Wisconsin Press.

Dupin, C. (2003), 'Early days of short film production at the British Film Institute: origins and evolution of the BFI Experimental Film Fund (1952–66)', *Journal of Media Practice*, Vol. 4, No. 2, pp. 77–91.

Geerhart, B. (June 2011), '*A Short Vision*: Ed Sullivan's Atomic Show Stopper', *CONELRAD ADJA-CENT* [online],

http://conelrad.blogspot.ca/2011/06/short-vision-ed-sullivans-atomic-show.html#_edn13, accessed December 2017.

Grant, M. (2013), 'Images of Survival, Stories of Destruction: Nuclear War on British Screens from 1945 to the Early 1960s', *Journal of British Cinema and Television*, Vol. 10, No. 1, pp. 7–26.

Gault, M. (2016), 'Watch the Nuclear Cartoon That Terrified Children in the '50s 'A Short Vision' traumatized Ed Sullivan's viewers', *War is Boring* [online], https://medium.com/war-is-boring/watch-the-nuclear-cartoon-that-terrified-children-in-the-50s-45ea4276a96e, accessed 9 January 2022.

Huntley, J. (1956), 'Animation in Britain', *Journal of the British Film Academy*, Winter 1956–57, pp. 4–6.

Jacobi, C. (2014). '"A Kind of Cold War Feeling" in British Art 1945–1952', in C. Jolivette(ed.), *British Art in the Nuclear* Age, London & New York, pp. 19–50.

Klinger, B. (1997), 'Film history terminable and interminable: recovering the past in reception studies', *Screen*, Vol. 38, No. 2, pp. 107–128.

Manvell, R. (1961), *The Living Screen: Background to the Film and Television*, London: Harrap.

Pitetti, C.M. (2016), *The City at the End of the World: Eschatology and Ecology in Twentieth- Century Science Fiction and Architecture*, PhD Thesis, The Graduate School, Stony Brook University: Stony Brook, NY.

Stephenson, R. (1973), *The Animated Film*. London: Tantivy Press.

Tai, P. (2013), 'The Aesthetics of Keyframe Animation: Labor, Early Development, and Peter Foldes', *Animation: An Interdisciplinary Journal*, Vol. 8, No. 2, pp. 111–129.

Taylor, H. (1989), *Scarlett's Women: Gone with the Wind and its Female Fans*, New Brunswick, NJ: Rutgers University Press.

Vasudevan, R. (2010), 'In the Centrifuge of History', *Cinema Journal*, Vol. 50, No. 1, pp. 135–140.

Part 2

Aliens: Allies and Adversaries

Chapter 3

Archetype as History in Hollywood Science Fiction of the 1950s

James Williamson

Introduction

With sudden force, Hollywood science fiction film exploded into a volatile existence in the early 1950s. Within the field of cultural criticism, these movies were traditionally regarded as somewhat disreputable, if not outright failures of the imagination, yet they played a role in creating an enduring niche within the film industry. Susan Sontag's critical examination of the genre in 'The Imagination of Disaster' is paradigmatic. She expresses some appreciation for the films, but concludes 'that the imagery of disaster in science fiction is above all the emblem of an inadequate response' to the atomic bomb, to the Holocaust, and to the Second World War (Sontag, 1961: 224). They were deemed inadequate by the industry too, for as the decade wore on budgets shrank and the genre began to acquire its reputation for cheapness and schlock, a recurring criticism from the end of the period onwards (see Amis 2012; Hodgens 1972). However, the very fact of the boom in sci-fi films – even the fact of their inadequacy – can still offer insight into the wider discussion of the relationship between individual films, their archetypal narrative forms and structures, and the cultural era which produced them.

Two early films of the genre that helped launch it to prominence, while demonstrating dramatically different approaches to the blending of mythic archetypes with the concerns of their historical period, are Robert Wise's *The Day the Earth Stood Still* (1951) and Byron Haskin's *The War of the Worlds* (1953). With their single cyclopean eyes, *The Day the Earth Stood Still*'s Gort and the Martians of *The War of the Worlds* appear to share a common iconic ancestry in some of the most famous mythic beasts of all, the Cyclopes of Homer's *Odyssey* (2003: 113–124). Where once such beasts lay beyond the sea, it is curious that at the dawn of space exploration some of these figures now descended from the sky as visitors. This was not always the case in the science fiction films in this period. Many of the 'beasts' featured in the 1950s corpus have more terrestrial origins, ranging from human science gone wrong, as in *Them!* (1954), *Tarantula!* (1955), and *The Wasp Woman* (1959), to creatures emerging from a mythic, primordial, and pre-human past after being disturbed by the encroachment of modern civilisation, as in *The Beast from 20,000 Fathoms* (1953), the original *Gojira* (1954), or *The Creature from the Black Lagoon* (1954). What distinguishes the 'beasts of the sky' in this period is not just their extra-terrestrial origin, but also a more distinct otherness, aliens in a true sense, while also tending towards being depicted as highly technologized, more sophisticated and advanced than mid-20th century humanity.

Yet this does not mean that visitors from outer space always carry with them the same significance. Both films do feature technological-yet-beastlike machines descending from the heavens, but with a very different purpose. They each make use of the archetypal and the historical, mediated by an emerging shared genre, however through their use they demonstrate fundamentally different approaches and values. *The Day the Earth Stood Still* is open-ended, questioning, with humanity's fate uncertain, *The War of the Worlds* returns humanity to what is supposed to be a more spiritually and ideologically 'purified' state of being – a Flood myth for the Atomic Age. Their approach influenced by the contingent effects of explicitly political concerns as well as narrative ones, these formulas are not rigid but fluid and dynamic. Not despite, but rather exactly because of, the formulaic repetitions and other 'inadequacies' of Hollywood science fiction film in this period, the films are in fact revealing of the relationships between the evolution of mythic archetypes and the expression of historical concerns.

The Day the Earth Stood Still (1951)

In archetypal terms, *The Day the Earth Stood Still* is primarily about a heavenly messenger, or rather, a pair of messengers – and the human response to their arrival. One presents as a genial white man in his late-thirties (a very efficacious appearance for an alien ambassador in 1950s America). The man is Klaatu, whose role is that of a kind of secular angel, and whose story draws its power, at least in part, from Christian resurrectionist mythology. His counterpart is the towering robot Gort, armed with sophisticated weaponry. Throughout the picture Gort is steadfastly, insistently present through his sheer size, yet he remains nearly featureless, silent, and inscrutable. His design is classically of the 1950s, emphasising smooth lines and barely sublimated phallic imagery. He is a threatening figure, but the precise nature of the threat varies throughout the film. In dispatching the soldiers set to guard him, he is cold and dispassionate, obliterating them without ceremony, remaining a true robot. Here we do see a modern, technologized breed of beast, precise and unfeeling as it kills, demonstrating no pleasure or remorse in the act. In this he specifically contrasts dramatically and drastically with his human – or, at least, more familiarly humanoid – companion. Where Gort is cold and removed, Klaatu is warm and personable; where Gort is silent, Klaatu is eloquent; and where Gort is destructive, Klaatu offers the promise, or at least the potential, of salvation. While it is Gort who serves as the film's threatening, beastly figure, and who provides the movie with its most striking contribution to the iconography of science fiction, his relationship to Klaatu is important for reading the robot's own significance within the narrative. They are partners, with a complex relationship of power and authority governing their behaviour.

The arc of Klaatu's character, his purpose within the story, can be parsed in relation to contemporary intellectual and academic interest in archetypal storytelling. For Hugh Ruppersberg, he is the ancestor of virtually all modern, cinematic iterations of the alien messiah, 'an expression of transcendence, from the first stage of vulnerability and closure to the second stage of transcendence and openness' (1990: 34–35). And for Northrop Frye, the messianic arc is itself the blueprint for the romance narrative: 'the hero of romance is analogous to the mythical Messiah or deliverer who comes from an upper world, and his enemy is analogous to the demonic power of a lower world' (1957: 187). Klaatu's quasi-preternatural status as an alien, as being literally from the heavens, and the purity which is associated with his utopian civilization, would seemingly mark him as being closer to the level of myth than the film's own opening rhetoric would have us believe. Unlike Frye's classical protagonist of romance, however, he has no single opponent in his quest to save the people of earth from themselves. The strange partnership he has with Gort, the film's 'monster', indeed emphasises that Klaatu is no two-fisted pulp hero of the 1930s and 1940s like a Flash Gordon or a Buck Rogers. He does not fight the beast, but rather brings the beast with him to earth. However, this turn, rather than detracting from, instead emphasises his quasi-messianic status. His enemy is not some being who represents the 'lower world': it is the lower world itself. It is the pettiness, greed, fear, and cynicism which he encounters from his first moment of stepping out of his interplanetary vessel. This is significant for the role Gort comes to play in the story.

Both Peter Biskind and Margaret Tarratt have observed the way the film's complex, even contradictory, attitudes towards the relationship between humanity and technology seem rarefied in the ambiguous figure of Gort. Biskind summarises Gort's role in the film with an emphasis on the primal destructiveness which he threatens:

> When Klaatu is harmed, Gort runs amok, and his first move after his master is "killed" is to stalk pretty Helen Benson. Gort and Helen are, of course, a replay of King Kong and Fay Wray with one crucial different ... King Kong, from darkest Africa, was the very avatar of a monster from the id if there ever was one, while smooth and shiny Gort, easily as dangerous, looks like it just dropped off the assembly line. Without the restraint of reason, technology is as dangerous as nature run wild. (Biskind, 1983: 157–158).

It is worth noting that just before Gort threatens Helen, he dispatches two soldiers with lethal force, where before he targeted only weapons and vehicles. There is a significant shift in Gort's behaviour here; he looms over her like a monster from a classic Universal Studios horror film, with the sexual threat which that implies. This supposedly logical machine leers at her with a literally harmful gaze. The camera emphasises his massive size and her vulnerability.

Even after she disarms him with the phrase Klaatu taught her (the famous science fiction shibboleth 'Klaatu barada nikto'), his sheer presence over-powers her, and he carries her inside the ship. For Tarratt, Gort "seems to represent man's violence and even his sexuality." Her suggestion is that the way the robot stands in opposition to everything the narrative praises about Klaatu is not incoherence, but is in fact the point: "The film suggests a concept of an ideal man separated from his most primitive instincts, using them only as a source of energy to aid his 'higher' civilized aims" (Tarratt, 2003: 356). Gort represents a technologized disassociation between 'man' and his basest impulses, allowing these beastly qualities to be controlled and directed towards a Universal Good, a triumph of self-mastery embodied by the apparently asexual, monk-like Klaatu. Beast and the monk separated, prose and passion divided; E. M. Forster's Margaret Schlegel would hardly approve (1999: 170). Taken together, these traits potentially mark Klaatu and Gort as figures of emotional – and, potentially, political – repression.

In this context, Klaatu's means of entry into the world from the heavens takes on an additional significance, one which has further resonance with the archetypal theories of the time. The key connection here is the emphasis Frye places on the importance of the literal or metaphorical birth of the hero as the first phase of romantic myth. "The first phase is the myth of the birth of the hero ... The infant hero is often placed in an ark or chest floating on the sea." Klaatu does, in fact, arrive on Earth by means of such a vessel – though he springs, Athena-like, fully formed into the world (in the hospital, he informs his human doctors that, despite appearing in his forties, his true age is seventy-eight). Of course, he comes not from the sea, as is typical for Frye's interpretation, but from the sky; from the heavens. These variations from the typical romantic structure are not random, however, but rather communicate specific meanings that still draw from a shared pool of arche-typal meaning.

As Frye notes of the image of the child in the sea, "Psychologically, this image is related to the embryo in the womb, the world of the unborn often being thought of as liquid." Klaatu arrives on Earth not through the material and messy process of a human birth, but by means that are, in archetypal terms, coded as divine and pure. Frye suggests this, noting the connection between the earth, the human soul, the celestial spheres, and a mediaeval Christian notion of deity:

> The symbolism of alchemy is apocalyptic imagery of the same type: the center of nature, the gold and jewels hidden in the earth, is eventually to be united

to its circumference in the sun, moon, and stars of the heavens; the center of the spiritual world, the soul of man, is united to its circumference in God. (Frye, 1957: 146).

They may not share the circular form of the sun, moon, and stars, but Klaatu's use of diamonds as currency echo this cosmic relationship, testifying to his status as the bridge between earth and heaven, as well as symbolizing once again his purity and innocence – not to mention standing as testament to the fabulous wealth of the utopian civilization to which he belongs.

The symbolism surrounding Klaatu and his behaviour marks him as a desexualised being of pure intention, any potential beastliness in him perhaps having been displaced onto Gort. Yet the nature of the supposedly utopian world, the ostensibly perfect and peaceful future which Klaatu and Gort together come to earth in order present and represent, therefore becomes deeply divided at its core. It combines elements of technological utopianism emerging out of Fordist scientific rationalism with revelatory millenarianism founded in Jewish and Christian myth and theology. Klaatu is both compassionate messiah, whose tolerance of humanity's flaws comes at the cost of his life, and accomplished scientist, with knowledge far outstripping even that of the film's Einstein-proxy, Barnhardt. The intersection of these two facets is further complicated by the film's idealisation of reason as a core virtue and its corresponding denunciation of certain forms of emotionality and corporeally governed experience.

Here we encounter one of the major criticisms of the film's message and its politics. The supposedly utopian, technocratic order Klaatu and Gort represent has been accused by Mark Jancovich not only of authoritarianism, but specifically of anticipating John von Neumann's formalisation of the concept of 'Mutual Assured Destruction', infamously known as MAD (Jancovich, 2004: 335). Certainly, the threat represented by Gort suggests a brutal and unforgiving universal order, but how this lines up with contemporary politics is complex. The ideology that gave rise to MAD is self-evidently *unilateralist*, insistent on portraying mid-century world politics as a binary state, a battle for supremacy between two competing power blocs with no nuance or middle ground (Hobsbawn, 1994: 229–230). Klaatu, by contrast, refuses to engage with the power plays of global politics, instead insisting on an unprecedented level of global co-operation that demands a surrendering of national pride towards a greater good, namely, the survival of all humanity. As such, he is closer to a *multilateralist* position than a unilateralist one. Ultimately, though, he remains above such territorial or even ideological concerns. As quasi-divine intercessor and judge, quite separate from any earthly arrangement of authority, Klaatu and Gort do not represent an order intended to be taken as a literal program that should (or even could) ever be enacted. The utopia they hail from and represent is ideal in the primary sense of functioning purely in the conceptual realm. Parsing their role in the narrative, then, requires attention to their symbolic and archetypal roles in order to access their contemporary political and historic significance.

The conflict between the scientific, rational view Klaatu espouses, and his actual role in the narrative as a religious, mythopoeic figure, is therefore no simple accident of incoherency, but part of a deliberate narrative and thematic suspension of, and an attempt to syncretise, these two knowledge systems. His dual nature is emblematic of the film itself. Thus, while one can dispute his emphasis, as Jerome Shapiro suggests *The Day the Earth Stood Still* is, in fact, something of a multi-valent hybrid: it is a biblical testament which nevertheless has "a structure that falls in line not only with ancient apocalypse narratives but also with what became in the fifties commonplace in bomb films." It is, in part, a messianic narrative where "the fulfilment of Biblical messianic prophecy starts in the liberal republic of science and then spreads to the larger society." It suggests the Jewish messiah as envisioned in the Psalms of Old Testament, but its concept of salvation is based upon the societal leadership of an intellectual elite who are instructed to promote moral action, and not on the internal spirituality which was the focus of ancient Judaism (Shapiro, 2002: 81–82).

The relationship between Klaatu and Gort represents a surrendering of power and control, of the ability to do violence, and is therefore not comparable to any terrestrial institution. The investment of such authority and institutionalised violence in Gort does not place him at the peak of the social and political structure of Klaatu's world(s), but rather, he is monstered by it. Gort's robotic-yet-beastly nature – his overlapping aspects of *in*human-ity – is precisely the point. Gort is presented as inhuman because the violence which it is his role to perform is inhuman. This is why his behaviour towards Helen is cast in a predatory light. Equally, it is important that the robot is neither quite subject nor object. Despite being identified as an arbiter of cosmic authority, Gort does not give, but receives verbal commands. Yet it is just as clear that Klaatu could not prevent Gort from fulfilling his purpose. The power relationship constructed between the robot and Klaatu is as inscrutable and mysterious as Gort himself, and this application of mystery is far from haphazard or lazy. It is a central part of the film's symbolic coding, an interweaving of archetypal narrative and imagery which is not just shaped, but defined, by the concerns of the historical moment.

The War of the Worlds (1953)

The relationship between historical expression and archetypal forms is neither stable nor consistent across the genre of science fiction film in the period. It is, essentially, inverted within the narrative of our other film, the George Pal-produced adaptation of *The War of the Worlds*. The framing and construction of this narrative occurs in a kind of contradiction to the cognitive shift in the cultural understanding of the causation of the apocalypse, that is to say the transition from the divinely ordained to the man-made. If, according to Daniel Wojcik, 'Apocalypse was now no longer a cosmic event executed by supernatural deities; it was now reduced to a mundane, technological absurdity' (1997: 103) then here we are confronted with a resistant narrative, an attempt to reassert the cosmic nature of apocalypse and return it to its properly ineffable place.

In his pointedly titled 1959 article 'A Brief, Tragical History of the Science Fiction Film,' Hodgens complained of the lack of imagination and ambition in 1950s sci-fi movies, though he praised Pal's attempts to expand the genre, arguing that 'if the scripts contained moments rather similar to more traditional spectacles, they still contained images that had never been seen before,' including *The War of the Worlds*' imagining of 'deadly Martian machines, like copper mantas and hooded cobras, gliding down empty streets' (Hodgens, 1972: 81). The following year, Kingsley Amis similarly lamented that 'it appears that the boom in science-fiction films has passed … and without having explored more than a fraction of the possibilities' (2012: 40). Like Hodgens, Amis despaired of the frequent resort to catastrophe and monster menaces, both foreshadowing Sontag's critical examination of the genre.

Yet Hodgens also takes care to note that '*The War of the Worlds*, with excellent Martians and some attempt to set up a logical alien technology, was probably the best of the menace series, if only because it provided a really formidable menace, one that couldn't be polished off with a few rounds of rifle fire' (2012: 40). The contemporary acknowledgement of a traditional approach to plot and storytelling framing a unique and distinctive spectacle – and, in particular, the rendering of fantastic technology through special affects to provide a genuinely intimidating foe for humanity – emphasises the combination of archetypal narrative forms and innovative cinematic technology, which opens up *The War of the Worlds* as a potentially rich site for the examination of the interaction between worlds old and new.

The Martians themselves are the attraction here, the source of the spectacle. Yet we are given little sense of the details or character of their civilisation or culture. This is not in spite, but exactly because, of the scale of the existential threat they pose. As Robert Markley suggests in *Dying Planet*, across their many depictions in popular culture 'Martians … are both the Other, the embodiment of an absolute intelligence that we can neither understand nor defeat, and those others who rival and mirror our desires and weaknesses and feed off of our own fears' (2009: 206). In *The War of the Worlds*, the most significant details we are given about these Martians are to do with their biology. As one of the movie's many scientist figures states: 'I don't remember ever seeing blood crystals as anaemic as these … They may be mental giants, but by our standards, physically they must be very primitive' (*The War of the Worlds*, 1953). These details, of course, set up the eventual revelation that the Martians cannot survive the microbiological fecundity of the living Earth, no more than their cyclopean eyes can tolerate light brighter than that of the Martian day.

Their biology therefore encapsulates their archetypal role as a force of enervation, their connection to the cosmic threat of entropy. Their encasement in formidable machines of war only emphasises the point – the modern science of cybernetics already inverted, twisted, and gone wrong. The archetypal portrayal of the Martians thus does not mean their representation is devoid of connections to specific contemporary concerns, or that these connections are merely superficial. The film's most remarkable special effects focus precisely on the destructive power of the Martian war machines, whether they are issuing a withering repudiation of the military's attempt at defence, or carving a burning path through the streets of Los Angeles – bringing a delirious nightmare of modern technological conflict home to America.

The mechanics of the Martian assault itself are significant in characterising how the archetype is being rendered here, indeed because, for all their chaotic, even demonic, role in the narrative, their methods are so distinctly mechanized and rationalised. The technologized approach of the Martians to warfare extends beyond merely a matter of their chosen tools. Their tactics are also representative of the role they play; for the Martian war machines, we are told, do not function as individual titanic warriors in automated armour, but as part of a larger network. This is made explicit in a scene where the leader of America's armed response to the alien threat, General Mann (Les Tramayne), briefs his audience of international dignitaries and military figures on the alien invaders' strategy:

> Now, this much is certain. It's vital to prevent the Martian machines from linking up. Once they do, they adopt an extraordinary military tactic. They form a crescent. They anchor it at one end and sweep on, until they've cleared a quadrant. Then they anchor the opposite end and reverse direction. They slash across country like scythes, wiping out everything that's trying to get away from them (*The War of the Worlds*, 1953).

The Martian war network, and the problems it presents to the united human military, bears some resemblance to the developments of the field of cybernetics as it emerged in the late 1940s and early 1950s, including the work of Norbert Wiener. In the perfectly synchronised combat strategy of the Martians there is also a resemblance to Wiener's concern that 'The danger of cybernetics … is that it can potentially annihilate the liberal subject as the locus of control' (Hayles, 1999: 110). The Martians are not only a threat to the individual on the level of physical annihilation the body and mind, but also in their mode of operation, their invasion force acting as a vast interconnected machine that stands in opposition to individual liberty.

Wiener, arguably the leading figure in the formation of cybernetics, had a particularly personal interest in the use and misuse of technology, as Hayles records: 'Wiener's war work, combined with his antimilitary stance after the war, illustrates with startling clarity how cybernetics functioned as a source of both intense pride and intense anxiety for him' (1999: 110). The ambivalence Hayles describes is of a different nature and tenor than that which inhabits science fiction film, but there is a significant correspondence in the duelling excitement and dread directed towards scientific and technological process, and the interface between the body and the machine. Wiener's cautioning against the misuse of technology does, in fact, take on an openly prophetic and apocalyptic tone: 'Like so many Gaderene swine, we have taken unto us the devils of the age, and the compulsion of scientific warfare is driving us pell-mell, head over heels into the ocean of our own destruction' (1989: 129). Anxiety over technology is, certainly, a very particular concern of the post-war period, as Wiener's project, and that of his fellow cyberneticists, demonstrates. Yet there is also a deeper, more elemental, anxiety in play here, which the specific act of projecting these anxieties on a nebulous, almost featureless, foe brings out.

The Martians are not only capable of destruction on a massive scale, but they are also destroyers of the bonds which cohere the human form and, even, those which construct meaning – whether these are the bonds between atoms or the social and spiritual bonds of matrimony and the church. As Sontag notes, 'These alien invaders practice a crime which is worse than murder. They do not simply kill the person. They obliterate him' (2009: 221). Martian technology corrupts and is corrupted, an extension of their nature as a twisted reflection of modern civilisation, invoking the use of nuclear weapons as well as the technologized and rationalised genocidal programmes that comprised the Holocaust, but here applied even more indiscriminately and rapaciously. This, for Markley, is characteristic of the role of Martians in science fiction invasion narratives:

> As the archetype of all remorseless invaders, "Martians" signify the dark, nightmarish underside of a modernist ideology that places its faith in science, technology, and progress. At the same time, they have served as stand-in for paranoid fears of malignant intelligence: Nazis, communists, body snatchers, fifth columnists, and all aliens who sap human strength and subvert identity from within (2005: 206).

The second point here is significant in the all-embracing nature of this form of xenophobia. Martians, considered *as* an archetype, do not denote a particular enemy, but are instead a particular idea of 'The Enemy', which can be detached and re-applied as suits the climate of the times. The Martians of *The War of the Worlds* noticeably lack a single direct link or analogical connection to a specific fear or anxiety belonging to their contemporary context.

Indeed, while *The War of the Worlds* may be a product of the Cold War environment, beyond the scale of the threat they pose, and the underlying notion of two worlds in conflict with each other, there is nothing necessarily to mark its Martians' behaviour as a representation of the Soviet Union specifically, despite the potential of the 'Red Planet' imagery. Indeed, with their blitzkrieg-like attack on civilization, they could just as easily be interpreted as spectres of the Nazi war machine, or Imperial Japan's sudden attack on Pearl Harbour. Markley's characterisation of the Martians as representing not only an alien Other but specifically 'those others who rival and mirror our desires and weaknesses and feed off of our own fears' (2005: 206) echoes Norbert Wiener's cautions that 'our present militaristic frame of mind ... has forced on us the problem of possible countermeasures to a new employment of these agencies on the part of an enemy. This enemy may be Russia at the present moment, but it is even more the reflection of ourselves in a mirage' (Wiener, 1989: 128). Here there is a connection to Wells's original narrative, which specifically draws a link between Martian invasion to European colonialism when the narrator wonders 'Are we such apostles of mercy as to complain if the Martians warred in the same spirit?' (Wells, 2010: 208–209). The evil that Martians do adapts and shifts with the times, evolving to match the fears which populate the imaginations of the specific era in which they manifest.

The Martians are, in fact, the embodiment of the same fears that lends Wiener's work its fatalistic tone. As he states in *The Human Use of Human Beings*, 'Progress imposes not only new possibilities for the future but new restrictions. It seems almost as if progress itself and our fight against the increase of entropy intrinsically must end in the downhill path from which we are trying to escape' (1989: 46–44). Indeed, Wiener's pessimistic diagnosis of humanity's progress recalls Walter Benjamin's famous reflection on Paul Klee's Angelus Novus: 'But a storm is blowing in from paradise; it has caught in his wings with such violence that the angel can no longer close them. The storm irresistibly propels him into the future while his back is turned, while the pile of debris before him grows skyward. This storm is what we call progress' (1999: 25). At the end of the path Wiener describes, in the wake of Benjamin's apocalyptic storm, is the dying planet Mars, its technological power belied by the physical weakness of its people. *The War of the Worlds* invokes this haunted understanding of technological progress, of the cost of advancement; but it is not truly interested in engaging with the implications of these ideas.

For while *The War of the Worlds* is a distinctly reactionary picture, this reaction is not expressed through the conveyance of a coherent political stance, though its politics are evident in the narrative's greater respect for the military and more regressive treatment of its female characters. The film offers a reassertion of the 'previous millenarian visions' that 'emphasized the imminent arrival of a redemptive new era,' in contrast to the 'increasingly pessimistic' beliefs 'stressing cataclysmic disaster' which became increasingly prevalent in the post-war period (Wojcik, 1997: 103). The beasts of *The War of the World* are truly and profoundly from outside our world, alien and Other, and in their demonic abjection, and even the serendipitous nature of their defeat, they avoid directly implicating humanity in any moral or ethical considerations regarding the development and use of science and technology in warfare. Informed though it may be by the specific techno-logical anxieties of the era, it deliberately turns away from politics, and its technical innovations, which so impressed Amis and Hodgens, are turned towards the service of clothing a fundamentally traditional story of disaster and survival.

Conclusion

There is a shared existential anxiety and dread to the visions portrayed in these films; indeed, this is at the centre of their appeal as narratives. This tension was at the centre of Sontag's original critique; that 'by means of images and sounds ... one can participate in the fantasy of living through one's own death and more, the death of cities, the destruction of humanity itself' (1961: 212). For her, such a fantasy is recurring and enduring, but even in a reactionary picture such as *The War of the Worlds*, the shaping of the archetypal forms underpinning that fantasy is necessarily framed by the recent experience of worldwide technologized warfare, of blitzkrieg and Pearl Harbour, of Auschwitz and Nagasaki. The Martian beasts are, then, ghosts of recent horrors, but it is no paradox that their power to occupy this position is derived precisely from their archetypal role. The very attempt to assert the continuity of Christian apocalyptic tradition is itself a veiled response to the same uncertainty and doubt towards the future that *The Day the Earth Stood Still* addressed directly. The final line of *The War of the Worlds* is, after all, a prayer – an act of faith, not hope: 'In this world and the next. Amen.' There is an element of cultural self-reinforcement here, even a mythic self-address, in the way the film asserts the ancient, incomprehensi-ble, and arbitrary nature of apocalyptic disaster through its depiction of a long-planned alien invasion.

The mythic and archetypal elements of *The War of the Worlds* are therefore not mere dressing, but deeply founded in its themes and in the structure and resolution of its plot, even as they provide the excuse for its displays of spectacle. In *The Day The Earth Stood Still*, by contrast, the interaction of the historic with the mythic produces a kind of secular parable, measuring choice and compromise in the deployment of phenomenal technological power. Here the technological beast from beyond the sky is revealed as the ultimate peacekeeper, a means by which human fallibility is tested and ultimately rendered impotent. The film's message can be read alternately as either pro- or anti- (or entirely indifferent to) disarmament, for or against Mutually Assured Destruction, because ultimately it is not a film whose action is primarily concerned with what we do with the bomb, but rather what we do with ourselves now it is in our possession. Fraught with particular significance for American culture and society following its deployment of atomic weapons in order to defend itself and project its power further abroad and into the future, *The Day the Earth Stood Still* expresses profound discontent with the recent past by projecting its anxieties onto the future.

Alternately, then, *The Day the Earth Stood Still* attempts to excavate and address the anxiety-laden foundations of the scenarios it confronts, while *The War of the Worlds* veils and ultimately retreats from that confrontation. The behaviours of the respective alien interlopers are entirely aligned with these approaches, each technological beast standing not only at the heart of these films' visual iconographies, but also their central themes. Both films therefore have their symbolic and thematic elements grounded in the weaving of archetypal constructs together with their active political and technological concerns. Yet their respective visions of the future diverge dramatically based upon their very different uses of the mythic and historic, resolving into conflicting paradigms of doubt and certainty. Confrontation with these beasts of the sky is a trial that is revealing not just for the characters and worlds of these films, but for the cinematic imagination of films and filmmakers, and their ability to confront an uncertain future.

Bibliography

Amis, Kingsley (2012) *New Maps of Hell*, London: Penguin.

The Beast from 20,000 Fathoms (1953), Directed by Eugène Lourié [DVD]. USA: Warner Bros. Pictures.

Benjamin, Walter (1999), 'Theses on the Philosophy of History', *Illuminations*, trans. Harry Zorn, New York: Schocken Books, pp. 245–255.

Biskind, Peter (2000) *Seeing is Believing: How Hollywood Taught Us to Stop Worrying and Love the Bomb*, New York: Owl Books.

The Creature from the Black Lagoon (1954), Directed by Jack Arnold [DVD]. USA: Universal Pictures.

The Day the Earth Stood Still (1951), Directed by Robert Wise [DVD]. USA: Twentieth Century Fox.

Forster, E. M. (1999) *Howards End*, New York: Modern Library.

Gojira (1954), Directed by Ishirô Honda [DVD], Japan: Toho Studios.

Hobsbawm, Eric (1995) *The Age of Extremes: The Short Twentieth Century 1914–1991*, London. Verso.

Hodgens, Richard, (1972), 'A Brief Tragical History of the Science Fiction Film', *Focus on the Science Fiction Film*, ed. William Johnson, Englewood Cliffs, NJ: Prentice-Hall, pp. 78–90.

Homer (2003), *The Odyssey*, trans. E. V. Rieu, rev. D. C. H. Rieu, London: Penguin.

Jancovich, Mark (2004) 'Re-examining the 1950s Invasion Narratives', *liquid metal: the science fiction film reader*, ed. Sean Redmond, London: Wallflower Press, pp. 325–336.

Lucanio, Patrick (1987) *Them or Us: Archetypal Interpretations of Fifties Alien Invasion Films*, Indianapolis: Indian University Press.

Markley, Robert (2009) *Dying Planet: Mars in Science and the Imagination*, Durham: Duke University Press.

Ruppersberg, Hugh (1990) 'The Alien Messiah', *Alien Zone: Cultural Theory and Contemporary Science Fiction Cinema*, ed. Annette Kuhn, London: Verso, pp. 32–38.

Shapiro, Jerome F. (2002), *Atomic Bomb Cinema: The Apocalyptic Imagination on Film*, London: Routledge.

Sobchack, Vivian (2004) *Screening Space: The American Science Fiction Film*, New Brunswick, NJ: Rutgers University Press.

Sontag, Susan (2009) 'The Imagination of Disaster', *Against Interpretation*, London: Penguin Books, pp. 209–225.

Tarantula! (1955), Directed by Jack Arnold [DVD]. USA: Universal Pictures.

Them! (1954), Directed by Gordon Douglas [DVD]. USA: Warner Bros. Pictures.

Tarratt, Margaret, 'Monsters from the Id' (2003) in *Film Genre Reader III*, ed. Barry Keith Grant, Austin: University of Texas Press, pp. 346–365.

The War of the Worlds (1953), Directed by Byron Haskin [DVD]. USA: Paramount Pictures.

Wells, H. G. (2010), *The War of the Worlds, Classic Collection*, London, Gollancz, pp. 203–364

The Wasp Woman (1959) Directed by Roger Corman and Jack Hill [DVD]. USA: Allied Artists Pictures.

Wiener, Norbert, *The Human Use of Human Beings* (1989), London: Free Association.

Wojcik, Daniel (1997) *The End of the World as We Know It: Faith, Fatalism and Apocalypse in America*, New York: New York University Press, 1997.

Chapter 4

The Quest for New Cinematic Grammar in Denis Villeneuve's *Arrival* (2016)

James Keyes

There are two well-documented current trends in the critical analysis of science fiction film. The first is genre criticism, the virtue of which, according to Annette Kuhn (1990) in her seminal book-length study *Alien Zone: Cultural Theory and Contemporary Science Fiction Cinema*, is an appreciation of cinema in its totality as a social ritual, in which the spectator is situated as a cog in the machine of the cinematic apparatus. The second is found in *Liquid Space: Science Fiction Film and Television in the Digital Age* (2017), in which Sean Redmond uses cognitive science to explore the ways in which the spectator encounters spectacular moments, fueled by the development of digital technologies, in science fiction film. Redmond's work calls for further investigation into the neurological dimensions of spectatorship and the science fiction genre, but Kuhn's paradigm is so persuasive that it has discouraged alternative approaches (Carruthers 2018; Hall 2016; Johnston 2011; Sontag 1966).

This essay mobilises Denis Villeneuve's *Arrival* (2016) to integrate these lines of enquiry and move past them, using the influential work of genetic epistemologist Jean Piaget (1952) to argue that *Arrival* conceptualises cinematic grammar in science fiction film as both liberating and limiting. It first illustrates that the import of *Arrival's* spatio-temporal configuration is irresolvable against the backdrop of genre criticism alone, calling for a reassessment of Kuhn's work and inviting Redmond's cognitive perspective to the film. While *Arrival* does traffic in the science fiction film's signifiers as plotted by Kuhn (that is, structural conventions, iconography, and ideology), this essay shows that the film's prolepsis (flashforward) more significantly reflects Villeneuve's preoccupation with the crucial relationship between diegetic time, cinematographic syntax, and the means by which images are coded in the science fiction genre. This essay then draws on Piaget's path-breaking model of cognitive development - assimilation, accommodation, and equilibration – to show that *Arrival* formally thematises the process of learning how to read the cinematic image through the linguist Louise Banks' (Amy Adams) shifting experience of space and time as a corollary to the acquisition of Earth's alien invaders' language. The result is that *Arrival* theorises the underlying psycho-cinematic system that constitutes the readability of its prolepsis as analepsis (flashback) and therefore calls attention to both the readable yet epistemologically porous nature of the image in science fiction film.

The utility of this approach is that it reconciles an alarmingly large gap in scholarship on the science fiction genre and contributes to a fast-growing branch of film studies known as 'neurocinematics' (Hasson et al. 2008, p.1) more generally, the latter exceeding the purview of genre criticism and therefore providing a platform for further long-term sustainable interventions.

Literature Review

This section's goal is to review the extant literature and, in the case of *Arrival*, show that evaluations of the film's structural, thematic, and aesthetic hallmarks in current commentaries strikes a chord with Kuhn's (1990) blueprint for the analysis of science fiction films. It primarily explicates the science fiction genre's potential to act as an ideological, reflective, repressive, intertextual, and observational vehicle (Kuhn 1990, pp. v–vi) through *Arrival's* narrative, editing, cinematography, and mise-en-scène. Attention is then called to the argument that Kuhn's paradigm forecloses novel avenues of enquiry, which in the case of *Arrival*, is to the detriment of much critical writing on the film. The following pages thus lay the foundations for a discussion of the value of the prolepsis and its relevance to models of film spectatorship.

At the beginning of *Arrival*, an extreme deep-focus long shot situates an alien spacecraft against the backdrop of rural Montana, establishing an ideological opposition resonating with Susan Sontag's (1966, p.209) observation that the first phase of the traditional science fiction film is marked by 'the arrival of the thing'. Despite the event's currency as a staple of Cold War science fiction, the trope's use in *Arrival* fails to qualify as an indicator, to borrow from Steffan Hantke (2010, p.146), of:

> the high degree of self-cannibalising and self-mythologising typical of American culture in general, manifested in a wave of remakes of material from all periods in cinema and an endless series of recombinations of successful ideas and formats.

Vindication for this claim is to be found in the fact that the protagonist, linguist Louise Banks, is recruited by Colonel G.T. Weber (Forest Whitaker), a senior U.S. Army officer, to investigate the spacecraft's inhabitants' - known as heptapods for their seven limbs - unintelligible language. Weber represents the narrative's chief obstacle to Banks' pacifist approach, routinely confronting her in a fashion that betrays his belligerent agenda. With recourse to films of this genre, it can be seen that Weber's approach is

not entirely unjustified. A rudimentary example is the fate of the astrono-
mers in *The War of the Worlds* (1953), who, following a good-willed attempt
to communicate with a Martian spacecraft on Earth, are vaporised by its ray
gun. *Independence Day* (1996) is another concise example, its extraterrestrials
obliterating the U.S. Bank Tower in Los Angeles and the well-wishing
humans atop in the first of myriad attacks on the planet. In this vein, Harvey
O'Brien (2012, p.1) points out, 'heroes do not seek out adventure, they
respond to dire necessity'. Thus the violence of those retaliating in the
science fiction genre is typically figured as a 'morally righteous and socially
redemptive' (O'Brien 2012, p.29) force. In *The War of the Worlds* and *Independence Day*, violence becomes retaliatory, a mechanism of self-preserva-
tion. Whereas in *Arrival*, the endeavour to establish an ethos of righteous
violence is stymied through Banks' and the heptapods' pacifism, subverting
what Thomas Elsaesser has termed the 'affirmative' (1975, p.281) ideology
of the classical Hollywood narrative and challenging the 'consequentialist
model' (ibid.) of cinematic storytelling more generally. Akin to O'Brien's
observation is the point that in a typical Hollywood film, the protagonist's
initiative galvanises the narrative's causal trajectory, resolving its contradic-
tions and pushing the plot to a resolution. The trajectory of *Arrival's*
narrative is causal, that is, consequential in time, but time is not instituted
sequentially, remaining covertly circular for the bulk of the film. An illus-
tration of this thesis at work is found in the opening scene. The camera tilts
to reveal a bottle of wine and two half-empty glasses on a table in a long shot.
"I used to think this was the beginning of your story," Banks ruminates by
way of a voice-over, the montage that follows implying that Banks is
addressing her deceased daughter, Hannah. "Memory is a strange thing,"
she continues, "it doesn't work like I thought it did." In a fashion formally
commensurate with Banks' comments, the final scene crucially reveals that
the opening scene's mise-en-scène is that of Banks' home on the night of
Hannah's conception. Crucially, after the plot has been resolved. Banks is
narrating in the past-tense a scene from the future, concealed by the scene
that follows. Time does not flow from A to B in a linear fashion in *Arrival*,
the fact that space-time is discombobulated mitigating the impulse to assess
the efficacy of action itself. Given that action is determined by tracking
transformations in space across time, one must be wary of a narrative that
destabilises the concept of linear time itself. While Sontag's binary is
established and one of Kuhn's frameworks shown to be applicable, Villeneuve does not embark on this well-trodden path, subverting this ideological dichotomy through an unorthodox formulation of space and time. Given
that this issue is at the kernel of this essay's argument, it has been attended
to in closer detail than Kuhn's other observations will be.

Referring to films that instrumentalise intertextuality as a means of speaking
to one's 'cultural capital of prior knowledge' and/or are a venue for the
notoriously porous concept of the 'postmodern sensibility,' (1990,
pp.177–79) Kuhn argues that the act of referencing calls attention to an

active network of relations and practices in cinema. The lakeshore by Banks' house, the site of Banks and Hannah's discussion about Hannah's father, recalls the opening scene of Andrei Tarkovsky's *Solaris* (*Solyaris*, 1972), a deep-focus long take tracking the protagonist, psychologist Kris Kelvin (Donatas Banionis), as he meanders along the shoreline of a lake. Kelvin's parents' dacha is anchored in the background, dull like the backdrop of the lakeshore in *Arrival*, both captured through a neutral density filter. The similarities are not only stylistic, but thematic. The heptapods' spacecraft resembles the monolith in Stanley Kubrick's *2001: A Space Odyssey* (1968), both articles affecting significant technological transformation in their diegeses. The former engenders the capacity for prescience (albeit through the heptapods' language), while the latter imparts vital evolutionary knowledge to prehistoric hominids, facilitating the production of nascent tools and weapons. The din of the U.S. Army's helicopters through Banks' house at the beginning of the film in conjunction with Villeneuve's use of low-key lighting echoes that of the scene in which a UFO terrorises Jillian Guiler (Melinda Dillon) and her son Barry (Cary Guffey) in Steven Spielberg's *Close Encounters of the Third Kind* (1977). The litany of the science fiction genre's symbols in *Arrival* highlights the fact that Villeneuve is aware of his film's semiotic scaffolding. Given the semantic freight borne by these symbols, in that they function as nuggets of cultural capital by tethering *Arrival* to a lineage of popular genre cinema, the film resonates with the Kuhnian framework for intertextual interpretation.

Reflecting on the film's self-reflexive nature in a more recent study (2018), Tijana Mamula contends that *Arrival's* mise-en-scène and cinematography, in their recapitulation of the medium's silent-era staples, call attention to the process of spectatorship itself. Banks, she points out (p.545), is routinely framed in a straight-on long shot before Abbott and Costello's - the monikers assigned to the dyad by Banks and Donnelly (Jeremy Renner) - screen-like interface in the heptapods' spacecraft. Others feature close-ups, the mobilisation of frontality for presenting communication between Banks and the heptapods, and Banks' dramatic gestures, all of which strike a chord with Tom Gunning's (1986, p.64) influential description of the self-reflexive 'cinema of attractions'. Further evidence for Mamula's argument is marshalled with recourse to the scene in which Banks removes her decontamination suit and places her bare hand on the spacecraft's screen-like interface to communicate with Abbott and Costello. Deploying her own cultural sign, that is, a gesture towards herself and Donnelly while saying the word "human," Banks begins a process of visual communication. The heptapods respond in kind, expelling an ink-like substance from their tentacles to form circular logograms on the interface. It is with the subjugation of verbal language to visual signifiers such as Abbott and Costello's logograms in *Arrival* that Villeneuve emphasises the film's inherently visual nature. Armed with her own knowledge of the mechanisms of human language, Banks proceeds to decode Heptapod. As Banks interprets logograms on a screen in a dark chamber in the heptapods' spacecraft, the spectator interprets *Arrival* in a film theatre, mirroring the act. Thus from a Kuhnian perspective, *Arrival* can be construed as a film about the cinematic apparatus.

The film's mise-en-scène, according to Gerry Canavan (2016, p.497), manifests 'a grim mood of horrified premeditation that still characterises U.S. culture in the time of Donald Trump'. What's true about Canavan's point, holding its political disposition at arm's length, is that the film's subject matter, thematic preoccupation, content, and structure can be construed to broadly reflect its culture's *zeitgeist*. Tensions escalate between the U.S. and China throughout the film. Moreover, despite Villeneuve's unconventional method of storytelling, *Arrival* preserves the traditional three-act structure, the conclusion resolving its disorder and producing equilibrium marked by a heterosexual coupling. The film is saturated in military and intelligence iconography reminiscent of the films produced in the wake of the September 11th attacks such as the *Bourne* trilogy (2002–2007), namely camouflage uniforms, insignias of rank, heavy-duty vehicles, networked screens, firearms and explosives, and so on. And *Arrival's* final scene, Canavan goes on (2016, p.497), suggests that the future is already ineluctably mapped out, leached with the structure of feeling that characterises the Anthropocene epoch in 21st century. It can be seen that *Arrival* can be construed as a venue for reflection on the culture that produced it.

Unlike readings that endeavour to unearth a text's reflective and ideological potential, Kuhn writes (1990, p.91), psychoanalytic readings have no concern with socio-political preoccupations. The role of psychoanalysis is to tease out subconsciously repressed meanings in texts, Barbara Creed's influential application of Julia Kristeva' work on abjection of femininity in *Alien* (1979) in *The Monstrous Feminine* (1993) being a well-known example. Kuhn goes on to note the science fiction genre's fascination with 'mysteries of conception at birth,' (1990, p.92) and one is struck, as Anne Carruthers is (2018, p.322), by Villeneuve's preoccupation with reproduction (*Maelstrom* [2000], *Enemy* [2013], *Arrival* [2016], and *Bladerunner 2049* [2018]), resonating with the psychoanalytic notion of the primal scene; the infant's separation from the world and interest in own origins. Villeneuve structurally elicits the spectator's curiosity in one scene by juxtaposing a medium straight-on shot of Banks at the U.S. Army's base in Montana with a close-up of Hannah encountering a horse in a stable with Banks. The depth of field is reduced in both shots, to say nothing of the fact that Banks and Hannah also occupy the same space in the frame. Villeneuve's adherence to the rule of thirds, the 180-degree axis of action, and the use of a graphic match prevents the eye's inclination to reorientate itself with the new shot's mise-en-scène, encouraging the spectator to consider the relationship between Banks' recent experience and this would-be maternal memory. Or as Anne Carruthers puts it (2018, p.325), 'the positioning of the viewer in relation to pregnancy and reproduction does two things, it encourages to read Louise as maternal, and reproduction and transient and cyclical'. As Carruthers correctly notes, juxtapositions of this character betray Villeneuve's preoccupation with the process of procreation. One is prompted to consider conception in *Arrival*, which is in essence, qua Christian Metz's (1982) work on psychoanalysis and cinema, arguably a rumination on their own origins. In light of the fact that cinema 'is sold on the spectator's expectations of pleasure or diversion' (Kuhn 1990, p.146), *Arrival* can be profitably understood as a spatio-temporal enactment of Freud's theory of the primal scene, from which the spectator derives pleasure. Even this superficial exploration should highlight the extent to which Villeneuve's work is germane to this complex body of theory as Kuhn suggests.

Thus far, this essay has demonstrated that *Arrival* exhibits a plethora of the science fiction genre's traditional structural, thematic, and aesthetic hallmarks, which has hitherto furnished its critical commentators with myriad opportunities to construe it as such. Given that Kuhn's paradigm is all-encompassing in relation to these staples, few, to this essay's knowledge, have successfully dredged *Arrival's* waters for novel critical treasures. The following section assesses this shortcoming and suggests why a cognitive exploration of the film might yield more fruitful results.

Gap in the Literature

Highlighting the ways in which *Arrival* exposes a cavity in Kuhn's paradigm, Villeneuve's preoccupation with the perception of cinematic representations of space and time speaks to broader discussions orbiting film, human vision, and the value of cinema as a critical tool, a concept that *Arrival* itself structurally invites. It contends that the bulk of scholarship on *Arrival*, in mobilising traditional analytical methodologies, fails to adequately account for its chief accomplishment, which is the thematisation of the simultaneously enabling and limiting nature of cinematic grammar.

In the film's climactic prolepsis, Banks consults a groundbreaking monograph on Heptapod, of which she is the author. This recollection allows her to communicate by way of a code word with the intransigent General Shang (Tzi Ma) of the Chinese People's Liberation Army in the present. The intervention results in the army's stand-down and a global war is prevented. In the same prolepsis Banks is seen delivering a lecture on Heptapod based on her book, implying its potential as a universal language. Taking up an ideological line of enquiry (Mamula 2018, p.544), Mamula argues that Villeneuve:

> is reluctant to envision a (...) cinematic model that might be capable of accommodating rather than eradicating the multiplicity of languages, both human and non, in existence today.

The film's treatment of Heptapod, Mamula argues (2018, p.548), is rooted in imperial soil, bolstering her thesis by drawing on Miriam Hansen's (1994) compelling expostulation of Jean Epstein and Béla Balázs' theory of silent-film grammar as a universal language. Yet this interpretation, sophisticated though it might be, calls attention to an interstice in Kuhn's critical framework. *Arrival's* anthropocentrism, according to Mamula (2018, p.550), warrants criticism insofar as it precludes non-human subjects. In its celebration of Heptapod's promising future as a diegetic Esperanto, the film fails to account for the value of cultural and ontological difference. But by way of rebuttal, it is important to note that Banks is the film's point of focalisation, which is to say that she is constituted by the conventions of narrative film as a signifier of diegetic navigation. Indeed, even Canavan's (2016, p.495) left-leaning political commentary acknowledges that 'we discover that the present tense and jumbled presentation of these events is not a literary convention but a registration of the way Louise now actually perceives time,' which vexes Mamula's diagnosis by virtue of the fact that anthropocentrism serves an integral functional role in Villeneuve's cinematic experiment. Focalisation, according to Silke Horstkotte and Nancy Pedri (2011, p.330), refers to 'the filtering of a story through a consciousness prior to and/or embedded within its narratorial mediation'. Banks' narrative is mediated to the spectator through her, not only foreclosing an objective presentation of the diegesis, but rinsing the cinematic medium through with her character. To give an example, the juxtaposition of Banks' voice-over and the long shot of the table in her home in the opening scene, accompanied by mournful non-diegetic music, implies that Banks has borne witness to this space in this particular time. Yet there's a paucity of evidence to support this claim. Villeneuve is not concealing the fact that the shot is presenting the future. This deception is achieved exclusively through the audio-visual contraposition, a product of Banks' prescience qua Heptapod. Canavan (2016, p.493) has noted the parallel between the heptapods' simultaneous perception of space and time in the past, present and future, bearing the implication that *Arrival's* opening scene presents a manifestation of Banks temporally palimpsestic impression of the diegesis. The fact that the spectator is not furnished with a chronological purchase on time is a result only attributable to Banks as the narrative's point of focalisation. Anthropocentric as this might be, Banks' consciousness is the lynchpin in Villeneuve's cinematic experiment. This valuable finding is made possible by only a cursory examination of focalisation, hinting that the potential cognitive science has to add to the sum of human knowledge on cinema remains untapped.

What is this experiment? In *Arrival*, Villeneuve calls attention to the fact that the interpretative scope of a shot or scene is contingent on its placement in relation to other shots in a film. This is hardly *terra incognita*, the Kuleshov experiment springing to mind. Familiarity with the codes and conventions of narrative film is also a prerequisite, as is the one's cognitive capacity to draw causal connections between shots, scenes, and sequences. Familiarity of this nature is a condition that must be satisfied in order for the spectator to misconstrue the value and function of any piece of film in a contingent context. *Arrival* evinces this insofar as it shows that a shot in its generic context precludes certain interpretations, but that it's the spectator's knowledge of genre, codes, and conventions that allows him/her to enable this familiar grammar's unfamiliarity. Thus the film's value lies is its contribution to one's understanding of the process of spectatorship, highlighting the cognitive triad – the forging of causal links between shots, an understanding of the principles of narrative cinema more generally, and knowledge of generic conventions - that constitutes science fiction film spectatorship. An initial explanation is in order. *Arrival* itself invites this reading in the scene following Banks' first encounter with Abbott and Costello in their spacecraft, in which she placates the bellicose Colonel Weber with a story about British explorer Captain James Cook's landing in Australia in the 18th century. Explaining that Cook's men misunderstood the Aboriginal Australians' word "kangaroo" to mean "I don't understand," Banks persuades Weber that her translation of Heptapod must proceed with caution. Convinced for the time being, Weber takes his leave. In confidence, Banks reveals to Donnelly that her story was fabricated, but that it proves her point nonetheless. A formal analysis of this scene in tandem with Banks' story expresses Villeneuve's intention. An establishing shot situates the spectator in a makeshift military base in Montana. A j-cut conceals the transition to a shot of the interior of a nearby tent. The dialogue between Weber and Banks is shot in the continuity style, Villeneuve using the shot-reverse shot technique, eyeline matches between cuts, and obeying the 180-degree and 30-degree rules. Medium close-ups and close-ups become more frequent as the drama, itself tempered by three-point lighting and the deployment of match-on-action cuts to sustain the scene's kineticism, intensifies, The word "kangaroo," like each shot and transition in a film, serves a function. Cinematic language, instrumental as it might be to communicating information relevant to a narrative, as Banks' story infers, is nonetheless constructed to a certain point. As Robert Ray (1985, p.33) put it, the classical Hollywood narrative's:

> habitual subordination of style to story encouraged the audience to assume the existence of an implied contract: at any moment in a movie, the audience was to be given the optimum vantage point on what was occurring on screen.

As far as Ray is concerned, the spectator implies the meaning of a shot, scene, or sequence based on assumptions about narrative film, genre, and cinematic storytelling, from knowledge of previous films. These assumptions produce expectations, which are then satisfied for the spectator's pleasure. Thus Villeneuve highlights on one hand the potential genre harbours to preclude certain interpretations of shots, scenes, and so on, based on their repeated use in previous films and one's own assumptions about the cinematic apparatus. On the other hand, the scene just examined indicates that cinematic grammar maintains a certain degree of interpretative flexibility even when its context has foreclosed certain readings. This is only the tip of the iceberg. The following section argues that Villeneuve's preoccupation with this threshold, that is, at which one's learned knowledge of the codes and conventions of narrative film meets more neurologically innate ways of interacting with the medium, is explored through the prolepsis.

Intervention

The prolepsis' placement throughout *Arrival* is the key to highlighting Villeneuve's cinematic enterprise. The argument presented in this section stems from the fact that the bulk of extant scholarship on *Arrival* recognises the prolepsis' presentation as analepsis yet neglects to question the means by which Villeneuve achieves this, instead funneling this information into Kuhnian interpretations (Carruthers 2018; Mamula 2018; Sutton 2018). This essay has already explicated Mamula's line of enquiry in this regard, though its findings have been far from conclusive. Through brief considera-tion of the work of David Sutton and Anne Carruthers, the foundations on which the benefit of Piaget's research can be built are laid. Like Mamula, Sutton mobilises a Kuhnian framework in his interpretation. Following the reflective route, his approach dwells on the 'cultural construction of tempo-ralities,' ruminating on film as a conduit through which the social mores of Western culture are articulated (Sutton 2018, p.7). Anne Carruthers' (2018) article in *Film-Philosophy*, while ultimately raising questions about the ethics of reproduction and thereby holding *Arrival* as a mirror to the world, adds a new element to the critical frame. Deploying Daniel Frampton's work to argue that *Arrival's* representation of 'pregnant embodiment is ultimately thrown into question by the temporal re-ordering of the film's narrative,' Carruthers says that the spectator remakes the film through *a-priori* and *a-posteriori* concepts (Carruthers 2019, p.322). The implication of this argu-ment is that *Arrival* is made legible by the spectator's predigested experience of narrative film and an innate ability to relate sequences of events to one another. The spectator must possess both if he/she is to recognise this re-ordering, which suggests that the process of ordering narrative film in the spectator's mind is built on a linear perception of time. What's striking is that Carruthers correctly notes that one misapprehends the film's prolep-sis as an analepsis, but fails to discuss the cinematic mechanisms facilitating this judgement (2018, pp.322, 331, 333). An oversight of this nature in a sophisticated analysis such as Carruthers' serves to underscore the fact that the bulk of criticism on *Arrival* is related to this misapprehension in one way or another (Canavan 2018; Carruthers 2018; Mamula 2018; Sutton 2018). Therein lies an important discovery.

Why does the spectator misinterpret *Arrival's* prolepsis as an analepsis? In light of the fact that the film is about – insofar as any narrative film can be *about* a single subject – the process of learning, Piaget's concept of the 'schema' is a useful heuristic when it comes to unlocking the film's secrets (Piaget 1952, p.7). Defined as 'a cohesive, repeatable action sequence pos-sessing component actions that are tightly interconnected and governed by a core meaning,' a schema is a dynamic cognitive building block that conditions one's reaction to an incoming stimulant (Piaget 1952, p.7). A unit of knowledge formed by past experience, a schema helps shepherd one through understanding and action. To give an example, a spectator might associate a Stetson hat with the Western genre. With this piece of knowledge, it would not be unreasonable to expect other symbols of the Western, that is, horses, tumbleweed, a standoff, and so on, from *Easy Rider* (1969), the protagonist (Dennis Hopper) of which wears a Stetson throughout. When this expectation goes unfulfilled, the spectator adjusts the schema related to the Stetson hat, which is ultimately one tethered to iconography and genre. Using this concept, this essay argues that incumbent on the success of Villeneuve's sleight of hand in *Arrival* is the spectator's possession of the following schema: (i) experience of narrative film insofar as he/she is familiar with the concept of cinematic storytelling; (ii) knowledge of the science fiction genre's conventions; (iii) the capacity to draw inferences based on perceptual phenomena. Villeneuve's unique use of the prolepsis calls attention to the brain's innate function to extract a legible narrative from a film. It also suggests that the verisimilitude of this narrative - in the case of the science fiction genre - hinges on an understanding of form and content manufactured by the society that produced the film. The fact that Piaget's three stages of cognitive development are evidenced in both Banks and the spectator through *Arrival's* temporal revelation bolsters this thesis. But first an explanation. The term 'assimilation' (Piaget 1952, p.407), Piaget's first stage, refers to the process of using an existing schema to process new knowledge, an example being a spectator bringing their knowledge of science fiction to bear on *Arrival* to draw inferences, make assumptions, and so on. This also relates to cinematic storytelling more generally, concerning causal links between shots. It is assumed, for instance, that the space-time represented in Shot A precedes the representation of Shot B's unless communicated otherwise. Piaget's second stage, 'accommodation' (p.408), occurs when one's understanding of a new phenomena is incommensurate with his/her current schema for that phenomena, subsequently requiring adjustment. 'Equilibration' (p.409), the final stage, is when one has assimilated the bulk of this new knowledge into an adjusted schema. Banks displays symptoms of assimilation when contacting Abbott and Costello in their spacecraft, using her knowledge of human languages to negotiate the heptapods' language. It's important to note with respect to genre that Banks' perception of time at this point in *Arrival* is linear, as is the spectator's. Equipped with her knowledge of human languages, Banks commences the arduous of translating Heptapod to English. Meanwhile, parallel to this action, the spectator implements his/her knowledge of the science fiction genre and cinematic storytelling to interpret the action, make assumptions, and temper expectations. It has already been demonstrated that at this point in the film, one has no cause to suspect Villeneuve of tampering with the temporal sequence of the shots. What's striking about Piagetian accommodation is that Banks' experience of space and time transforms as a corollary to the acquisition of the heptapods' language, which, the spectator learns, implies an experience of circular time in its articulation. Banks' reality, that is, the reality experienced by Banks as the spectator's point of focalisation, becomes increasingly interrupted by scenes presented as flashbacks of her terminally ill child. As Banks' schema adjusts to the concept of circular time implied in the heptapods' language, in other words, so does the spectator's. Yet the spectator, David Sutton correctly points out, is not aware of it. 'As viewers,' he writes, 'we are enculturated into the grammar, or language of film, which [...] allows us to interpret the meaning of certain scenes, given our expectation of the conventions of the flashback' (Sutton 2018, p.9). One's familiarity with the science fiction genre's conventions and narrative film, that is, the expectation that significant lapses in time and the direction in which time is flowing in the diegesis are clearly signposted, in addition to the brain's capacity to draw causal inferences based on perceptual data, which then feeds back into Piagetian schemas of narrative film and genre convention, allows Villeneuve to violate the traditional contract between time and narrative without the spectator's knowledge. In one scene, for instance, the spectator sees Banks' receive news of her daughter's death, a j-cut accompanying Banks' voiceover between the fade out and fade in to another scene. The cut implies a linear transition from past to present between the two and invites the spectator to question the impact of this trauma on Banks' life, rather than the nature of the temporal shift itself. As this essay mentioned earlier, a j-cut also marks the linear transition between a shot of a military base and Banks and Weber's argument. This is not to suggest that the spectator is conscious of the transition. An awareness of the mechanisms of continuity editing undermines their efficacy. Nor is it this essay's intention to argue for the subconscious effect of the j-cut either. Has one exhausted all modes of critical inquiry such that the cognitive processes employed in film spectatorship must be understood only with recourse to the subconscious? This essay would argue otherwise. Another example will help support this claim. Andras Kovacs' (2007, 2011) work, which builds on the back of decades of theory and empirical studies, shows that a spectator infers a causal connection between any two juxtaposed shots in a film. The aim of continuity editing is to ensure that this connection is pertinent to a given film's narrative; linear time is implied in the word "continue" in that the filmmaker wants the narrative to proceed forwards in space and time. In one scene, a sound bridge connects a medium straight-on shot of Banks looking at the ground to a reverse shot of a mise-en-scène of lakeshore in subdued colour, which is then followed by a series of slow motion shots and fleeting images of Banks' daughter. Villeneuve uses the sound of Banks' boots on gravel in the present to motivate a cut to Banks and Hannah standing on a pebble-laden lakeshore. Once again, a j-cut facilitates this transition, Hannah's voice being heard before the cut to the lakeshore. The reverse shot is also crucial. Maintaining consistent spatio-temporal relations throughout the scene, the shot formally insinuates that Banks is once again slipping into her past with the aid of an aural trigger. In a medium close-up, she stares pensively at the lake. A reading of these conventions in the context of the science fiction genre leaves no doubt in the spectator's mind that Banks is reflecting on the past. For Banks, the past has yet to come to pass. For the spectator, this past is also communicated by the conventions of narrative film in the past relative to the present. The shot cuts, returning along the 180-degree axis to Banks, who is struck by moment of realisation. "I just realised why my husband left me," she says. With the belief that Banks as been married before, the spectator dispenses with the thought of a potential temporal rupture. As Jan Simmons has insightfully argued, the pleasure derived from a narrative of this nature stems from the revelation of the 'underlying process of construing a story and the elements out of which is is composed' (Simmons 2008, p.113). Shot through a wide-open aperture, the scene is overexposed and possesses an oneiric quality. The spectator believes that Banks' past is the source of the film's enigma, but Villeneuve has a different agenda. It is not until Banks asks one of the heptapods: "I don't understand. Who is this child?" that the spectator questions the nature of Villeneuve's representation of time. Thus characterised is the transition from Piagetian accommodation to equilibration in both Banks and the spectator. The representation of space and time, Villeneuve reminds one, is not a run-of-the-mill process. This is not to say that *Arrival's* representation of space-time does not express, to some degree, the spectator's relationship to it. Implicit in the process of recognising is the passage of linear time. Without space, occurence itself is foreclosed. It is with occurence that time is made manifest, and space is always already transforming as a result of passing time, incessantly resulting in new space; a new arrangement of matter persisting in time. The experience of linear time is in-built to the spectator's brain, a critical schema brought to bear on narrative film that *Arrival* exposes with the aim of encouraging serious consideration of the cognitive processes of spectatorship. This marriage of cognition and cinematic convention allows Villeneuve to execute *Arrival's* structural sleight of hand. Through Banks' experience of the prolepsis, the spectator is compelled to question the nature of narrative, the use of the science fiction film's socially produced aesthetic and structural codes, and reflect on the simultaneously enabling and limiting nature of the brain's capacity to draw causal links between shots and sequences, resolving in equilibration.

This essay has demonstrated that *Arrival* highlights a gap in Kuhn's seminal theoretical scaffolding on the science fiction genre, which has been unattended to thus far. It has suggested a cognitive approach, not as a means of supplanting this work, but as a means of circumscribing elements that Kuhn has been unable to appropriately address, for further fruitful critical scrutiny. Rather than arguing for *Arrival's* advocation of a universal cinematic grammar, this essay has underscored the ways in which Villeneuve suggests precisely the opposite. With the work of Jean Piaget, it has instead explored the idea that the prolepsis is deployed to call attention to the brain's innate capacity to construct epistemologically digestible narratives, but that the viability of these narratives in the case of the science fiction film rely partly on socially produced stylistic and structural conventions. Ultimately, it has revealed that the success of Villeneuve's sleight of hand in *Arrival* is the spectator's possession of the following Piagetian schema: (i) experience of narrative film insofar as he/she is familiar with the concept of cinematic storytelling; (ii) knowledge of the science fiction genre's conventions; (iii) the capacity to draw inferences based on perceptual phenomena. These findings should pave the way for further fine-tuned critical debates on the science fiction genre and film spectatorship more generally.

Bibliography

Canavan, G. (2018), 'Living in the Future'. *Science Fiction Film and Television*, 11 (3), pp.491–497.

Carruthers, A. (2018), 'Temporality, Reproduction, and the Not-Yet in Denis Villeneuve's *Arrival*'. *Film-Philosophy*, 22 (3), pp.321–339.

Creed, B. (1993), *The Monstrous Feminine: Film, Feminism, Psychoanalysis*, London: Routledge.

Elsaesser, T. (2004), 'The Pathos of Failure: American Films in the 70s', in Horwath, A et al. (eds.) *The Last Great American Picture Show: New Hollywood Cinema in the 1970s*, Amsterdam: Amsterdam University Press, pp.279-292.

Gunning, T. (1986), 'The Cinema of Attractions'. *Wide Angle*, 8 (3), pp.63–70.

Hall, J. 'Time-Travelling Image: Gilles Deleuze on Science-Fiction Film'. *Journal of Aesthetics and Education*, 50 (4), pp.31–44.

Hansen, M. (1994), *Babel and Babylon: Spectatorship in American Silent Film*, Massachusetts: Harvard University Press.

Hantke, S. (2010), 'Bush's America and the Return of Cold War Science Fiction: Alien Invasion in *Invasion, Threshold*, and *Surface*'. *The Journal of Popular Film and Television*, 38 (3), pp.143–151.

Hasson, U et al. (2008), 'Neurocinematics: The Neuroscience of Film'. *Projections: The Journal of Movies and Mind*, 2 (1), pp.1–26.

Horstkotte, S & Pedri, S. (2011), 'Focalisation in Graphic Narrative'. *Narrative*, 19 (3), pp.330–357.

Johnston, K. (2013), *Science Fiction Film: A Critical Introduction*, Oxford: Berg Publishers.

Kovacs, A. (2007), 'Things That Come After Another'. *New Review of Film and Television Studies*, 5 (2), *n.pag.*

Kovacs, A. (2011), 'Causal Understanding and Narration'. *Projections: The Journal of Movies and Mind*, 5 (1), pp.51–68

Kuhn, A. (ed.) (1990), *Alien Zone: Cultural Theory and Contemporary Science Fiction Cinema*, London: Verso.

Metz, C. (1977), *The Imaginary Signifier: Psychoanalysis and Cinema*. Translated by Britton C et al. Bloomington: Indiana University Press.

Mamula, T. (2018), 'Denis Villeneuve, Film Theorist; or, Cinema's Arrival in a Multilingual World'. *Screen*, 59 (4), pp.542–551.

O'Brien. H. (2012), *Action Movies: The Cinema of Striking Back*, New York: Wallflower Press.

Piaget. J. (1952), *The Origins of Intelligence in Children*. Translated by Cook, M. New York: International University Press.

Redmond, S. (2017), *Liquid Space: Science Fiction Film and Television in the Digital Age,* London: Bloomsbury.

Robert, R. (1985), *A Certain Tendency of the Hollywood Cinema,* New Jersey: Princeton University Press.

Simmons, Jan. (2008), 'Complex Narratives'. *New Review of Film and Television Studies*, 6 (5), pp.111–126.

Sontag, S. (1966), *Against Interpretation and other Essays*, New York: Farrar, Straus and Giroux.

Sutton, D. (2018), '*Arrival*: Anthropology in Hollywood'. *Anthropology Today*, vol. 34, no. 1, pp.7–10.

Part 3

Animating Sky-Borne Beasts

Chapter 5

The Magic and the Mundane in *Avatar: The Last Airbender* and *The Legend of Korra*

Francis M. Agnoli

Where does a sky bison poop?

The world of the *Avatar* television franchise is brimming with life. The forests, seas, and deserts overflow with fantastical fauna, invented animals distinct from those of the real world. Within the sky, great beasts soar in physics-defying flight, instilling awe in both human characters and human viewers alike. Such creatures appear throughout the animated television series *Avatar: The Last Airbender* (2005–2008) and its sequel *The Legend of Korra* (2012–2014), revealing aspects of the fluctuating dynamic between humans and nonhuman animals, as well as between magic and the mundane.

There is a long history of depictions of animals in media. Those who study such texts emphasize how these images and narratives provide insight into real-world human-animal relationships (Molloy, 2011, p. 1; Burt, 2002, p. 15; Rothfels, 2002, p. vii). Paul Wells comes to the same conclusion regarding his chosen field of study: "Animation is a ready vehicle to both illustrate and exemplify the relationship between animals and humankind through the necessity and consequences of change" (2008, p. 134). The fluidity or plasmaticity of animated figures is a long-established trait of the medium, with Sergei Eisenstein observing their "rejection of one-and-forever allotted form" (1988, p. 21). More recently, both Wells and Jane Batkin extrapolate that fluid forms yield fluid meaning (Wells, 1998, pp. 118, 213; Batkin, 2017, pp. 1–2). These animated animals are teeming with oscillating and sometimes contradictory commentaries about human-animal relationships that viewers can unpack and relate to everyday life. Furthermore, as Wells notes, the medium of animation affords artists greater control over an image than live-action filmmaking practices would (Wells, 2002, p. 73). A filmed animal may be coached, tricked, or edited together to produce the desired performance, but there remains the potential for serendipity and spontaneity. The animated animal is entirely within the control of artists. Even their voices are often provided by human actors, such as Dee Bradley Baker for the *Avatar* television franchise. Fantasy introduces additional distancing. Kathryn Hume argues that all fiction is composed of the dual impulses toward mimesis and fantasy, toward the imitation of and the breaking away from reality (1984, p. 20). Within this dynamic, realistic elements ground fantastical ones (Irwin, 1977, pp. 9, 189; Worley, 2005, p. 14). As fantasy narratives, *Avatar* and *Korra* balance magical and mundane elements in their world-building, including in their invented fauna.

With few exceptions, the animals of the *Avatar* world are hybrids of real-world referents (e.g., otter-penguins, platypus-bears, wolf-bats, and turtle-ducks). They are distinct from actual animals with which viewers would be familiar, yet a hyper-realist aesthetic serves as a grounding agent. As defined by Wells, hyper-realism refers to how some animated texts approximate filmed representations of reality (1998, p. 25). If these creatures really existed, then this is how they would look and behave if recorded. However, they do not exist – they are conceptually impossible, even if their anatomy and behavior resemble live-action equivalents. Because these animated animals do not represent specific real-world counterparts, they are able to stand-in for whichever species or groups of species the artist or viewer chooses. Therefore, the rendering and depictions of these fantasy creatures comment on human-animal relationships. For the majority of human characters in these shows, animals exist primarily as sources of food, clothing, labor, transportation, companionship, entertainment, annoyance, or danger – much as real-world humans view nonhuman animals. There are only a few exceptions.

As prominent beasts of the sky, sky bison (alternatively called air bison) and dragons complicate these dynamics, both between human and nonhuman as well as between magic and the mundane. Due to these creatures' seemingly impossible ability to fly and their scarcity, human characters are frequently overcome with awe when they encounter them. In contrast, other animals – while likewise fantastical to the viewer – are regarded as ordinary. At the same time, sky bison and dragons often serve practical functions. They act as companions for the Avatars as well as steeds within the Air Nomads and the Fire Nation, respectively. These fluctuating fictional and fantastic relationships are illustrative of fluid real-world human-animal relationships, as we observe an overall shift from seeing these creatures as magical, awe-inspiring beings with interiority to treating them as exploitable resources defined by their mundane utility.

Sky Bison

Over the course of two shows, the *Avatar* world undergoes industrialization and modernization. These changes are reflected in how both the series and the human characters regard the more fantastical elements of the show, from bending, to the Avatar, to sky bison. As the balance shifts from magical to mundane, these creatures transform from sources of awe to mere beasts of burden. Featured in the first series, Appa is the most prominent sky bison of this franchise. In addition to being Avatar Aang's animal companion, he serves as the principal means of transportation for the young and often broke heroes. Like the eponymous Avatar and last airbender, Appa is the last of his kind and representative of all sky bison. In order to understand what this creature says about the relationships between human and nonhuman animals as well as between magic and the mundane, this section reviews Appa's appearance and behavior across key episodes. These observations are then compared to the treatment of his species in the sequel series.

While all sky bison combine characteristics of manatees and bison, they are not straightforward hybrids. As co-creator and art director Bryan Konietzko writes, "this is just how bisons [sic]... appeared in our fantasy world" (DiMartino and Konietzko, 2010, p. 80). Sky bison are large, horned beasts covered in white fur; brown markings in the shape of an arrow run along their back. When grounded, they walk on six legs. When in flight, they occasionally flap their flat tails and swim through the air like manatees (DiMartino and Konietzko, 2010, pp. 20–21). The minimal movement allows for the cost-saving recycling of frames. Such mechanics are impossible and should situate this creature firmly within the realm of fantasy. Nevertheless, because this animated animal is rendered through the same methods as the human characters – through two-dimensional animation – he is fully integrated into their world, is presented as a natural part of the fauna. There is already an interplay of magic and the mundane. These elements inform how Appa behaves, how human characters regard the sky bison, as well as how individual episodes frame him.

Appa's status as a source of awe is reflected in how different human characters react upon seeing him flight. When he first ascends in "The Avatar Returns" (2005), Sokka – the most cynical member of the cast – responds with joyous wonder. Later in the episode, when Prince Zuko spots the creature, he stops fighting and stares in disbelief. In "Imprisoned" (2005), a group of Fire Nation guards struggle to describe the creature to their commanding officer. Such reactions are not typically extended to other creatures, regardless of their enormity or abilities. Appa's large size and uniqueness also present logistical challenges for both the human characters and the episode writers. Often, when Aang and his friends visit a new town or are traveling incognito, Appa becomes a liability and is left on the outskirts. He is shooed away for the majority of the episodes "The King of Omashu" (2005), "The Headband" (2007), and "The Painted Lady" (2007). Sometimes, his temporary absence results in a heroic last-minute rescue, as seen in "The Waterbending Scroll" (2005), "The Deserter" (2005), and "Return to Omashu" (2006). Such recurring narratives may raise questions regarding Appa's activities when he is not interacting with his human companions. What does he do when left alone for long stretches of time? Was he anxious when they did not return for a day or longer? Was he bored? How did he react when he heard Aang summon him?

Sometimes, the episode writers address these subjects, giving Appa his own b-story or subplot. In "The Cave of Two Lovers" (2006), the heroes are trapped underground. As usual, the narrative focuses on the human characters. During key scenes, Appa is either off-screen or standing unmoving in the background. Again, the animators conserve time and money by recycling frames. However, there are moments when Appa leaps to the foreground, displaying some sense of personality and desire outside of the needs of the human characters. The sky bison does not like being underground, and he regularly expresses his discomfort and anxiety throughout the episode. After the initial cave-in, he frantically tries to dig himself out before resigning with annoyed growls. When the humans think they found a sealed exit, Appa charges forward unprompted, breaking down the stone door and almost running over Aang and Katara. When they finally escape, the sky bison rushes out ahead of his group, flops onto his back, and exhales. The storyboarding of these actions clearly conveys Appa's deep sense of relief. In "The Swamp" (2006), Appa and the winged lemur Momo are separated from their humans in the titular ecosystem, granting them long stretches of uninterrupted screen time. Again, Appa displays his individuality and personality in his reactions to the hostile environment. He repeatedly tries to fly out and gets trapped in the vines, collapses in frustration with the impassable terrain, and roars at the noisy nocturnal fauna. He also recognizes that members of the native Foggy Swamp Tribe are hunting him for food and quickly swims away. Through these scenes, Appa is shown as more than his mundane utility, as more than a mere means of transportation or source of companionship for the human characters. He is also presented as more than an awe-inspiring spectacle. He is an individual with a distinct personality and interior life, whose existence extends beyond his relationship with humans. Such moments add up across episodes; as noted by television scholars, the format can harness repetition and variation to create richer characterization (Fiske, 1987, pp. 84, 108; Burns and Thompson, 1989, p. 3; Corner, 1999, p. 57). Eventually, Appa can carry his own episode.

"Appa's Lost Days" (2006) depicts a four-week period when the sky bison is captured and separated from the human protagonists. Over the course of the episode, the writers, character designers, storyboard artists, and animators depict a range of actions, emotions, and appearances. Appa fights against his captors and a territorial boar-q-pine. He sneaks food behind the back of a circus trainer. He reminisces about his childhood. He is shown as ferocious, mischievous, fearful, hungry, nostalgic, confused, restless, and at peace. Although Appa's character design does not afford much leeway for depicting facial expressions, the storyboard artists are able to convey the sky bison's emotional reactions through head tilts and framing. In addition, character designers Angela Song Mueller and Jae Woo Kim produced a range of special poses and costume (SPC) designs for this episode (Di-Martino and Konietzko, 2010, p. 118). Appa is shown in a circus outfit, as battle damaged, and as a young calf. All of these elements unite in a commentary on positive and negative human-animal relationships.

Throughout this episode, Appa suffers from interactions with humans who do not view him as an individual with interiority but instead as a resource to be exploited. The sandbenders and beetle-headed merchants treat him as a commodity to be sold or traded. The circus sees him as a lucrative source of entertainment. Azula and Long Feng use him as a tool to capture or control Aang. Even a farmer and his wife see him as some*thing* to fear rather than recognize him as some*one* who is scared. In contrast, Appa has positive interactions with humans who acknowledge his suffering and attempt to alleviate it. A child at the circus sees that Appa is hungry and sneaks him food, Suki and the Kyoshi Warriors feed and wash him after the battle with the boar-q-pine, and Guru Pathik provides sanctuary at the Eastern Air Temple. The latter two parties also respectfully give Appa distance until he relaxes to their presence. Through these encounters, "Appa's Lost Days" communicates how humans should interact with animals and how they should not. While the narrative and subject remain fantastical, the writers did tether this episode to real-world concerns, drawing parallels with the real-life abuse and exploitation of circus animals. The Humane Society of the United States cited this resonance when they awarded this episode with a Genesis Award in 2007. In fact, the choice of centering such a narrative around a fictional creature allows his story to apply to a range of non-fictional ones. Just as Appa is an individual rather than a prop or part of the background, so too are all real-world animals. Reducing them to sources of meat, entertainment, or threat – as various antagonists across these episodes do to Appa – is perverse. Furthermore, all nonhuman animals possess interior lives that humans should acknowledge and respect. However, within this franchise, such privileges can be conditional.

Set seventy years later, *Korra* takes place in a world undergoing modernization and industrialization. This new series has fewer episodes per season, a more serialized narrative structure, and a larger cast of recurring human characters. Due to these factors, less screen time is devoted to the primary animals companions of the series – Naga the polar bear dog and Pabu the fire ferret. Secondary animals, such as members of newly discovered herds of sky bison, receive even less development. All of these changes impact the depiction of this beast of the sky, as seen in their appearances, behaviors, and relationships with humans. In both form and design, these new sky bison emulate and evoke Appa. They are still large, six-legged creatures that swim through the air. They are still rendered through the same animation methods and techniques as the human characters. Only the slightly altered markings on their back suggest that they might be a different breed of sky bison than Appa. There are few variations between individuals. Avatar Korra's unnamed bison from the second season has a broken horn, Kai's bison Lefty is a bit shaggy, and Opal's bison Juicy is a mess of matted hair and snot. However, these are exceptions. The introduction of multiple sky bison could have led to a greater emphasis on their individuality, presenting a plethora of unique personalities and distinct characterizations. Instead, their interchangeable external appearances reflect an apparent lack of internal psychology, as these nonhuman characters are stripped of their ability to inspire awe and are reduced to their mundane utility.

Oogi – the steed of Tenzin, son of Aang and Katara – is the most prominent sky bison in the sequel series. Yet, he lacks any discernible interior life. In "Welcome to Republic City" (2012), he flies Tenzin and his family to meet with Korra. The purpose of this scene is not to introduce Oogi, who is not referred to by name until later in the episode, but instead to spotlight the new human cast. After landing, the sky bison exists immobile in the background or is off-screen entirely. He does not interact with any of the human characters beyond his function as a mode of transportation. This trend continues in subsequent appearances, such as in "When Extremes Meet" (2012), when Korra and Tenzin converse while on their way to Air Temple Island. Although the scene opens and closes with wide shots of the sky bison in flight, the framing focuses on Tenzin and Korra's tête-à-tête. Even though he is rendered through the same techniques as the human characters, the animated animal is reduced to being part of the background. This relegation recurs with other sky bison, such as Lefty in "After All These Years" (2014). Like Oogi, Lefty is featured in establishing shots before the scene cuts to medium shots and close-ups of the humans. He operates as an unmoving part of the background, barely interacting with anyone else. Neither the farmers nor the villagers react to his presence. While Appa may have occasionally been figuratively and literally relegated to the background, such scenes did not constitute the majority of his appearances. Now that sky bison are commonplace, they are no longer special. They are not sources of awe but instead are mundane modes of transportation, not individuals with interiority but interchangeable beasts of burdens.

Sky bison receive the spotlight again in the *Korra* episode "Original Airbenders" (2014). The narrative is still human-centric, focusing on Tenzin teaching the new generation of airbenders and on his daughter Jinora learning to be more assertive. While the episode opens with domesticated bison serving utilitarian functions, the new airbenders shortly thereafter express excitement at the sight of a wild herd. The script explicitly draws parallels between the airbenders and the sky bison. In the final scene, as the humans watch the calves take flight, Tenzin intones: "I guess everyone is growing up." Like "Appa's Lost Days," this episode offers contrasting human-animal relationships. Earth Kingdom poachers reduce sky bison to sources of food and clothing, with their leader wearing a baby bison pelt as a cape. The episode frames these tendencies as evil and perverse. The trading and consumption of bison steaks – as well as the meat of all exotic and endangered animals – is condemned by the sympathetic humans. Again, the episode draws parallels to the real world. Initially, the wild sky bison perceive all humans – both poachers and airbenders – as threats, with a new mother charging at Kai to protect her children. It is only after the airbenders rescue the calves from the poachers that the adult bison relax their guard. Observing this change in behavior, Tenzin explains: "The bison are the original airbenders. They recognize their own kind." This interspecies relationship is framed as a sacred bond. However, such consideration is not widely extended beyond this episode. The special connections between individual airbenders and sky bison are not developed, certainly never to the extent as Aang and Appa's. The sacred airbender-bison bond is instead taken for granted, as with Kai and Lefty, or is joked about, as with Opal and Juicy. Airbenders may not eat these creatures, but they do regularly reduce them to means of transportation.

In the first series, the depiction of Appa offers an idealized human-animal relationship, one that recognizes and respects the consciousness of the nonhuman, frequently contrasted with exploitative human behavior. However, that human-sky bison dynamic shifts over time. When an animal ceases to be special – such as when they are no longer the last of their kind – then it becomes easier to reject or ignore their individuality and interiority. Unlike Appa, the sky bison of *Korra* are no longer unique within this fantasy world. Therefore, they are no longer sources of awe and spectacle but instead are principally viewed as modes of transportation. They possess little identity beyond that of a loyal and passive steed, and the new generation of airbenders rarely treat them as anything else. However, the sky bison are not the only beasts of the sky in this franchise. Another species – one adapted from cultural referents rather than real-world animals – offers additional insight into the oscillating relationship between humans and nonhuman animals as well as between magic and the mundane. Here, there be dragons.

Dragons

Like their fellow beasts of the sky, the dragons of *Avatar* and *Korra* have a fluid relationship with humanity, one marked by a steady transition from magical to mundane in the face of modernization and industrialization. Dragons make fewer appearances across the two series than sky bison, and these examples can be divided into two categories: the physical and the metaphysical. The former refers to those who exist as part of the natural fauna of this fantasy world. The latter are otherworldly, belonging to the lands of dreams and spirits, taking on symbolic weight. Together, both types illustrate the scopes of these depictions and what they say about human-animal relationship.

Unlike sky bison, dragons are not original creations for this franchise. Instead, *Avatar* and *Korra* draw on a long history of depicting these mythical creatures. On one hand, the writers and artists emulate European traditions, ones that relate to the majority of their cultural backgrounds as well as those of the assumed North American audience. At the risk of flattening a range of diverse cultures and their mythologies, the dragons of Medieval Europe share key traits in appearance and function. They are reptilian creatures with vicious claws, bat-like wings, and fiery breath (Fee, 2011, p. 7). As Jack Zipes describes them, dragons are "terrifying beasts" akin to lions and wild boar within European fairy tales (2007, p. 5). Narratives featuring these animals focus on heroic dragonslayers, such as Beowulf, Saint George, and Siegfried (Fee, 2011, pp. 3–26). Dragons themselves often take on demonic or evil connotations in antagonistic roles (Zhao, 1992, pp. 52–56; Fee, 2011, p. 7; Honegger, 2019, p. 117). On the other hand, crewmembers of *Avatar* and *Korra* also found inspiration in Chinese traditions, which informed the bulk of the franchise's world-building. Again, at the risk of flattening millennia of myths, legends, and folktales, the Chinese dragon or *long* stands in stark contrast from their European cousins. While there are many different types of *long*, most are depicted with horns, serpentine bodies, and whiskers (Zhao, 1992, p. 19–20). Some but not all are wingless. Whereas the European dragon breathes fire, the *long* is typically associated with water (Zhao, 1992, p. 61; Yang, An and Turner, 2005, pp. 101–102). While there have been Chinese folktales featuring malevolent *nielong* and heroic would-be dragonslayers (Zhao, 1992, pp. 6–7, 10; Yang, An and Turner, 2005, p. 105), the *long* are predominantly depicted as guardians and helpers delivering divine aid (Wilson, 1990, p. 299; Zhao, 1992, p. 6; Yang, An and Turner, 2005, p. 100). This combination of European and Chinese referents informs the nature of human-dragon relationships in *Avatar* and *Korra*.

In *Avatar*, dragons have been hunted to the brink of extinction. As a result, the physical variants are rarely depicted. Fang is the first dragon introduced in the series. As the animal companion of Roku, the previous Avatar, his posthumous appearances are limited to the Spirit World and flashbacks. Next, Ran and Shaw live in hiding among the Sun Warrior ruins, where they teach ancient firebending techniques to worthy humans. Finally, there is Druk, Zuko's steed in the third season of *Korra*. Each of these depictions – in their design, behavior, and framing – offer insight into human-animal relationships as well as how the interplay between magic and the mundane impacts that dynamic.

The designs and animation of these four physical dragons demonstrate a melding of European and Chinese traditions. Fang establishes the template. His long, serpentine body is covered with red scales. Two horns extend back from his head, and two long pink tendrils snake out from either side of his mouth. Four short legs keep his belly low to the ground. When in flight, Fang sporadically uses his large, bat-like wings, as his sinuous body quickly slithers through the air. Designed by Jae Wood Kim and colored by Hye Jung Kim, Ran and Shaw are far larger than Fang, even though they copy much of his appearance (DiMartino and Konietzko, 2010, p. 115). One duplicates Fang's crimson motif while the other has blue scales. Both have furry beards and protruding fangs, giving them a more savage mien. When in flight, they move like eels through water, changing direction with balletic speed and grace. They use their leathery wings when hovering in place. Finally, as designed by Angela Song Mueller and colored by Sylvia Filcak-Blackwolf, Druk possesses a somewhat more muscular appearance than his antecedents (DiMartino, Konietzko and Dos Santos, 2015, p. 34). Rather than a continuous, ophidian body, Druk's torso is clearly segmented from his neck and tail, although certain angles in later appearances do suggest a more snakelike build. Otherwise, the design as well as flight animation repeats established elements. All of these dragons are rendered through the same animation methods as the human characters, Despite their apparent otherworldliness, they are indeed part of the natural world. Physically, these dragons are emulative of the Chinese *long*, with their horns, tendril-like whiskers, and serpentine bodies with short legs. However, these traits are combined with European ones, including prominent chiropteran wings and, most notably, fiery breath. The melding of European and Chinese inspirations can also be found in the behavior and framing of these beasts of the sky.

Fang is introduced in "Winter Solstice, Part 1: The Spirit World" (2005). Aang is stuck in the Spirit World when he sees a dragon flying toward him. He panics and tries to flee. After a commercial break – indicating that this creature is a cliffhanger-worthy threat – Fang lands in front of Aang and wordlessly reveals his identity as Roku's animal companion. The dragon then guides the young Avatar through the Spirit World, showing him how to contact Roku and how to return to the physical world. Aang – and, by extension, the viewer – first see Fang as a malevolent threat only to learn that he is a source of divine aid. At the risk of essentializing, this moment can be read at an initial European or Western instinct giving way to a Chinese or Eastern understanding of this animated animal. In "The Fire-bending Masters" (2008), Aang and Zuko's investigations into the origins of firebending lead them to Ran and Shaw. The two humans are surprised to learn these masters are actually a pair of dragons, the last of their kind. In this scene, the two beasts of the sky are literally larger-than-life. The frame rumbles as they flap their wings and circle around Aang and Zuko. How else could humans respond to such a spectacle except with fear and awe? These are highly intelligent creatures who communicate their secrets through actions rather than through words. Aang and Zuko recognize a pattern in the dragons' flight, and they begin to copy their movements, learning an ancient firebending technique. Ran and Shaw then pass judgment on the humans before showing them the true meaning of firebending via a colorful spire of flame. In light of such intelligence and wisdom, the notion of hunting these creatures is depicted and perverse and shameful. It is another sign of how spiritually and ecologically unbalanced the imperialistic and mechanized Fire Nation has become since the time of Avatar Roku.

When introduced in "Rebirth" (2014), the dragon Druk is framed as equally if not more important than his rider, an elderly Zuko. The shot follows the former prince and an unnamed guard through a hallway before pushing past them, revealing Druk posed regally against the dusk. The episode ends shortly thereafter, with the dragon and his rider flying toward the horizon. The storyboarding, animation, and lighting instill a sense of awe for this beast of the sky, the first physical dragon depicted in this franchise since Ran and Shaw. A scene from a later episode mines this type of reaction for humor. In "Long Live the Queen" (2014), a group of Earth Kingdom soldiers return from the desert and see Druk. After a stunned pause, the commanding officer suggests that they all go get drinks. However, outside of these moments, Druk is regarded by both the show and the other characters in a more mundane manner. He is first and foremost Zuko's steed. This attitude typified in a scene from "The Ultimatum" (2014), where Korra and Zuko have a conversation while the latter is saddling Druk. Throughout a series of shots, the dragon functions as a static background for the two humans, with his face occasionally visible. Only his intermittent blinks are animated. The scene's seeming disregard for Druk mirrors attitudes toward the sky bison throughout the sequel series. Here, though, such treatment is contrasted with the initial reverence. Even though *Avatar* dragons are consistently framed as sources of wisdom and awe, that attitude is tempered in *Korra*.

While Fang, Ran, Shaw, and Druk all belong to the fauna of *Avatar*, dragons in this television franchise are also depicted as otherworldly. Even Fang, while originally of the natural order, ultimately ascends to the Spirit World, a privilege never afforded to Appa in the sequel series. Otherwise, metaphysical dragons make two major appearances. First are the unnamed blue and red dragons from Zuko's fever dream in the first series. Second is the Spirit Dragon from an extended flashback in *Korra*. While their scenes are brief, these dragons once again reveal aspects of the human-animal relationship, largely through how they combine European and Chinese referents.

The character designs for these dragons foreground their Chinese inspirations. Designed by Jae Woo Kim and colored by Hye Jung Kim, the dream dragons from *Avatar* are even more snakelike in appearance than Fang (DiMartino and Konietzko, 2010, p. 121). Their limbless, wingless bodies coil around the pillars of the Fire Lord's throne. Although they do not fly, these dragons still move with an apparent weightlessness. They have tufts of hair on their chins and large fins growing out of the side of their heads, emphasizing their species aquatic associations. As with Ran and Shaw, one is red and the other is blue. The Spirit Dragon from *Korra*, designed by Il-Kwang Kim and colored by Sylvia Filcak-Blackworld, stands in contrast to all previous dragons (DiMartino, Konietzko and Dos Santos, 2015, p. 93). Their white, wingless, and anfractuous body flies through the air like an eel would swim through water. The more overtly Chinese visual elements seem to correspond with more mystical behavior, framing, and relationships with humans. Despite their explicit ethereal nature, these metaphysical dragons are still rendered through the same animation techniques as human and nonhuman characters.

In "The Earth King" (2006), Zuko falls into a fever dream and has a vision of two dragons fighting over his soul. The blue one speaks with the voice of his sister, lulling him to sleep. The red one speaks with the voice of his uncle, urging him to leave. Here, these metaphysical dragons take on a pronounced symbolic weight. In the real world, the *long* often represents the Chinese emperor and as well as China as a whole, and the designs and framing of the *Avatar* dragons tether them to that context (Wilson, 1990, p. 305; Zhao, 1992, p. 6; Yang, An and Turner, 2005, p. 100). For this scene, dragons operate as metaphors for the Fire Nation, specifically its dueling malevolent and noble sides. The same battle takes place within Zuko, torn between the influences of his sister and of his uncle. This significance is then deepened through Zuko's subsequent interactions with physical dragons, both Ran and Shaw in *Avatar* as well as Druk in *Korra*, as the prince redeems himself and his nation. In both instances, Zuko is paired with a red dragon, revealing which of the two metaphysical entities he heeded.

Similarly, the Spirit Dragon scene from *Korra* establishes an ancient connection between dragons and firebenders. Set 10,000 years before the sequel series, the episodes "Beginnings, Part 1" (2013) and "Beginnings, Part 2" (2013) follow Avatar Wan's journey through the Spirit Wilds as he learns how to bend the four elements. In the first episode, the Spirit Dragon teaches him firebending. The choreography and framing intentionally parallel the tutelage of Ran and Shaw. According to in-universe mythology, the dragons were the original firebenders, just as the sky bison were the original airbenders. Humans learned how to manipulate these elements by emulating these creatures. These origins stories mirror those of many real-world martial arts, which also cite the movement of specific animals as inspiration (Farrer, 2013). This brief scene depicts a harmonious human-animal relationship, one where the former can learn from the latter. It is a type of camaraderie that could only exist in this ancient era, before the physical variants of such creatures were reduced to big game to be hunted or to steeds to be ridden.

Like the sky bison, the dragons of *Avatar* offer insights into human-animal relationships. Across their few appearances, these beasts of the sky operate primarily as sources of awe, as both spectacles and as providers of divine wisdom and aid. The hunting of them to the brink of extinction is framed as perverse, as a sin for which both Zuko and this nation must atone. The metaphysical manifestations of these creatures deepen their symbolic importance, tying them more closely to the Fire Nation and to firebenders.

Unlike the sky bison – a wholly original creation – these dragons meld certain cultural referents in their appearance, behavior, and framing. The Chinese referents signify the ancient and mystical nature of these beasts of the sky, whereas the European ones address more mundane and pragmatic concerns. After all, how can these animals fly without wings?

Conclusion

Media depictions of animals, including the animated beasts of the sky in *Avatar* and *Korra* offer insight into human-animal relationships. These series often provide idealized versions of that dynamic. Good humans respect and protect sky bison and dragons; bad ones reduce them to their basic utilities. Viewers can apply these lessons to their daily lives, with the episode writers drawing explicit parallels to the real-world abuse and exploitation of nonhuman animals by circuses, poachers, and big game hunters. However, as part of a long-running television franchise, *Avatar* and *Korra* also depict these relationships as in flux. In the preindustrial societies of *Avatar* and the "Beginnings" two-parter in *Korra*, the human protagonists and the narratives still recognize these nonhuman animals as individuals with unique personalities, agendas, and interiority. In the postindustrial societies of the sequel series, such creatures are more easily commodified as beasts of burden, even though the poaching of sky bison is still condemned. At the same time, *Korra* offers hope about the restoration of functionally extinct species. The introduction of herds of sky bison and of Druk provide a utopian resolution to the narratives of genocide and ecological devastation from the previous series, even if they seem to coincide with the loss of individuality. By *Korra*, the human characters have largely stopped thinking of these beasts of the sky as beings with interiority or as sources of divine wisdom. They no longer wonder how they are feeling or what lessons they can impart. Instead, they are left with more pragmatic concerns. How fast can they fly? How many people can they carry? Where does a sky bison poop?

Bibliography

Batkin, J. (2017). *Identity in Animation: A Journey of Self, Difference, Culture, and the Body*. London: Routledge.

Burns, G. and Thompson, R.J. (1989). "Introduction" in Burns, G. and Thompson, R.J., eds., *Television Studies: Textual Analysis*. New York: Praeger, pp. 1–6.

Burt, J. (2002). *Animals in Film*. London: Reaktion Books.

Corner, J. (1999). *Critical Ideas in Television Studies*. Oxford, UK: Clarendon Press.

DiMartino, M.D., and Konietzko, B. (2010). *Avatar: The Last Airbender – The Art of the Animated Series*. Milwaukee, OR: Dark Horse Books.

DiMartino, M.D., Konietzko, B. and Dos Santos, J. (2015). *The Legend of Korra: The Art of the Animated Series – Book Three: Change*. Milwaukee, OR: Dark Horse Books.

Eisenstein, S. (1988). *Eisenstein on Disney*. Translated from Russian by Alan Upchurch. London: Methuen.

Farrer, D.S. (2013). "Becoming-animal in the Chinese Martial Arts" in Dransart, P., ed., *Living Beings: Perspectives on Interspecies Engagements*. London: Bloomsburg, pp. 145–165

Fee, C.R. (2011). *Mythology in the Middle Ages: Heroic Tales of Monsters, Magic, and Might*. Santa Barbara, CA: Praeger.

Fiske, J. (1987). *Television Culture*. London: Methuen.

Honegger, T. (2019). *Introducing the Medieval Dragon*. Cardiff, UK: University of Wales Press.

Hume, K. (1984). *Fantasy and Mimesis: Responses to Reality in Western Literature*. New York: Methuen.

Irwin, W.R. (1977). *Game of the Impossible: A Rhetoric of Fantasy*. Urbana, IL: University of Illinois Press.

Molloy, C. (2011). *Popular Media and Animals*. Houndmills, UK: Palgrave Macmillan.

Rothfels, N. (2002). "Introduction" in Rothfels, N., ed., *Representing Animals*. Bloomington, IN: Indiana University Press. pp. vii-xv.

Wells, P. (1998). *Understanding Animation*. London: Routledge.

—. (2002). *Animation: Genre and Authorship*. London: Wallflower.

—. (2008). *The Animated Bestiary: Animals, Cartoons, and Culture*. New Brunswick, NJ: Rutgers University Press.

Wilson, J.K. (1990). "Powerful Form and Potent Symbol: The Dragon in Asia," *The Bulletin of the Cleveland Museum of Art*, 77(8), pp. 286–323.

Worley, A. (2005). *Empires of Imagination A Critical Survey of Fantasy Cinema from Georges Méliès to* The Lord of the Rings. Jefferson, NC: McFarland and Co.

Yang, L., An, D. and Turner, J.A. (2005). *Handbook of Chinese Mythology*. Oxford, UK: Oxford University Press.

Zhao, Q. (1992). *A Study of Dragons, East and West*. New York: Peter Lang.

Zipes, J. (2007). *When Dreams Come True: Classical Fairy Tales and Their Traditions*. New York: Routledge.

Yoichiro Kawaguchi's (1987) *Ecology II: Float II*

Chapter 6

From Ocean to Outer Space: Digital Creatures Surviving in Generative Animation and the Re-evolution of Games

Chunning (Maggie) Guo

The *French Lieutenant's Woman* used an open ended narrative device, offering readers three different possible paths for the book's ending. This kind of interactive technique for the development of effective entertainment is not new. Filmmakers have long relied on test audiences to ensure the popularity and impact of motion pictures. Shifting the role of audiences towards that of creative authorities has also inspired the development of a generative system to create "digital creatures" automatically.

Since the 1980s, computers have been central to the development of new artistic works and even new digital lives. Pioneering generative animation artists have transformed the experiments of narrative structures into digital evolution. The first generation of these animated works ventured to explore mysterious creatures in the ocean. From tracing models of marine organisms (*Aquarelles*) to 3D construction of digital creatures (*Growth Model*), generative animation opened up new possibilities for discussing the topic of evolution. Generative animation provides an opportunity to weigh the debate of creationism and evolution, which also expands a large space for imagination of the diversity of the ecosystem in outer space. Generative art is partly supported by the theory of genetic systems. Digital genetic systems are different from the real world. Through interactive behavior with audiences, these digital creatures from the ocean to outer space are experiencing an adventure of re-evolution. Especially in some generation animation installations such as *Galapagos* by Karl Sims, audience members or participants are playing a God-like role, determining which creatures have the right to survive.

Re-evolution is not only the theme of some generation animation, it is also becoming a new interface. The entire process of making a choice, creating a new creature and enjoying adventure in a video game could be regarded as a testament of re-evolution. Such as in the game *Spore*, created by Will

Wright, which consists of a series of behaviors of choosing, constructing and destroying. In *Spore*, players themselves create an entire universe, advancing from a single-celled creature in the water to an inter-galactic species.

Through vital abstract visual principles, *Spore* presents a new, evolution-like process, which this paper calls "re-evolution", as this is a highly personal-ized, humorous and cooperative adventure. Each choice by the player greatly affects the development of their own cultivated creatures' development, and each time a new game starts, the creatures are different from the previous ones, a metaphor for parallel universes. Players, in effect, construct the digital revolution together with each of their choices contributing to the development of their creatures and determining whether the fate of their planet will be survival or invasion from other alien species. The game also accommodates more tangible preferences in that players may approach problems using diplomacy or violence as they see fit.

Re-evolution as the interface of generative animation is also connecting to cross- disciplinary research. In the book *A Legacy of Freedom* by French geneticist Albert Jacquard, the author describes a unique species of human beings who, through the ability of free choice, break the barriers of the genome and their environment. The choices of Heterogeneity demonstrated by a variety of generative animation works resonate with global scientific genetic experiments. These artistic explorations also raise questions related to scientific ethics with implications for decision-makers, particularly in the domain of genetic research and editing.

Preface: The Choice of Heterogeneity as a Bridge

Since the late 1980s, the open structure became a focus for novel writers and film directors. It is not only about a series of experiments on art form, it also reveals the desire of rethinking historical and political systems. Especially following with the development of the personal computer, the open struc-ture was involved in more installation art, animation, as well as digital games, contributing to the generative system creation. Whether it be open structure or a generative system both related the shifting authority from author to the reader, audience or player. The readers have more freedom in making choice for the ending of a novel, the audience has plenty of choices on storyline, and the players could remake their own adventure in each new round of a game.

This paper defines this kind of choice making as the choice of hetero-geneity. The concept of Heterogeneity is on the opposite side of the spec-trum from homo-geneity. Heterogeneity was first used in the field of physics, referring to the multiform of materials and space (spatial hetero-geneity). This concept was also borrowed in biology and genetics, referring more to behavior, such as French geneticist Albert Jacquard who raised his theory about the freedom in gene choice and established the unique charac-

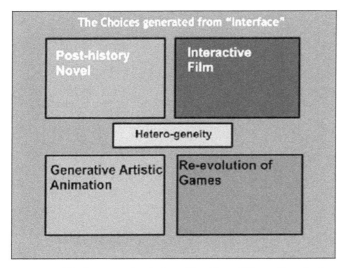

Figure 1: The heterogeneity choices offered from different "interface"

teristics of human beings.[1] Heterogeneity in sociology is more related with a continual negativity, especially about the opening theory of art and anti-mainstream broadcasting. So in artistic practices, Heterogeneity was presented as avant-garde deconstruction and new system experimentation.

This paper will examine the diversity of choices offered by four types of "interface", which are post-history novel, interactive film, generative artistic animation and re-evolution of games (see Figure 1).

Furthermore, this paper will interpret how these multiple choices can be regarded as an animated process. Thus, the "interface" is not only a physical in-between medium, but also a fantasy of re-evolution, full of new choices, votes and creations. Nowadays, the human beings are already attached with these attractive interfaces by unlimited touching, choosing and playing. The result of "living on" the interface is actually a re-evolution of the human being, turning themselves into a new species, which is "interface-being".

To witness this process of re-evolution, we need to trace back to the late 1960s to examine the open-ending of post-story as a process of "animated choices", and interpret the experimented narration of interactive film, as a metaphor about voting behavior just an illusion of democracy. This paper will also demonstrate how the digital interface became a new generative world, the digital creatures from the ocean to the outer space rehearsing a game of re-evolution. However, behind the choice of Heterogeneity offered from interface, we will also experience the limitation of freedom, or the limitation of human being. Though more choices can be made in the system based on gene theory and global genetic scientific experiments ultimately

1   Albert Jacquard, *L'Héritage de la liberté de l'animalité à l'humanitude*, Editions du Seuil, 1986. (Chinese translated edition), translated by Gong Huimin, GuangXi Normal University Press, 2005 July. p. 8.

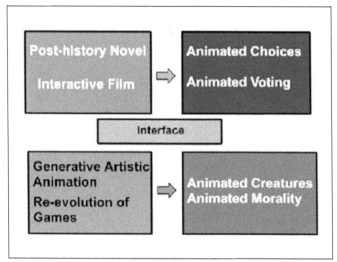

Figure 2: Multiple choices connected with the creation and behaviors in animation.

bring new challenges, especially gene ethics to the choice makers.

Post-history Novel with Open Endings

The post-historical novel *The French Lieutenant's Woman*[2] written by John Fowles was published in 1969, this novel included open structure offering creative choices to readers. Canadian scholar Linda Hutcheon describes the novel as an exemplar of a particular postmodern genre: "historiographic metafiction."[3] On the surface this is a fraught relationship of an amateur naturalist Charles Smithson and an independent Sarah Woodruff, while underneath it demonstrates the ambitious goal of rewriting history, or the author believes that there is only written history. In this novel, the romantic adventure of Charles' pursuit of Sarah leads to three possible endings:

A: Charles' marriage to his previous fiancée Ernestina (not Sarah).

B: Charles' successful reestablishment of a relationship with Sarah.

C: Charles is cast back into the world without a partner.

These are not the only options of a romantic relationship between a gentleman and a formal female governess, it is also related to the multi-perspectives of how people can explore history. Michelle Phillips Buchberger discusses these endings as a demonstration of "Fowles's rejection of a narrow mimesis" of reality or history,[4] Fowles presents this multiplicity of endings

2   John Fowles, *The French Lieutenant's Woman* (London, 1969).

3   Linda Hutcheon, *A Poetics of Postmodernism: History, Theory, Fiction* (New York, 1988).

 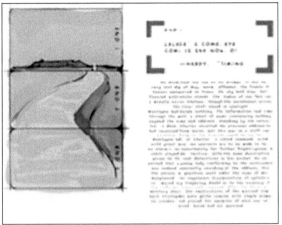

Figure 3: The multiple endings of the novel *The French Lieutenant's Woman* gave inspirations to a book of design. WangXiao Yu (王晓宇), the student in Art School Renmin, University of China, interpreted the novel as a design of multiple choices divided into three envelopes in the reimagined edition of *The French Lieutenant's Woman*.

to highlight the role of the author in plot choices. In this novel, readers may wonder if this French Lieutenant which Sarah has fallen in love with really exists. This illusion about the French Lieutenant also refers to the idea that "There is no such thing as history, there is only written history" (inspired from E.H. Gombrich).[5] The multiple endings of the novel *The French Lieutenant's Woman* gave inspiration to designers and artists for new book designs and filmmaking (see Figures 3 and 4).

Figure 4: *The French Lieutenant's Women* is also an ideal example to rethink the image of women. Liu HuiZh (刘慧至), a student of the Art School Renmin University of China, redesigned the film poster of *The French Lieutenant's Women* by reconstructing three combinations of the images using photos of her own hands.

4   Michelle Phillips Buchberger, *John Fowles's Novels of the 1950s and 1960s. The Yearbook of English Studies*. Modern Humanities Research Association. 42:133. 2012.

5   Ernst H. Gombrich, *The Story of Art* (London, 1995), p. 5.

Voting Choices for Interactive Film

In 1981, *The French Lieutenant's Women* was successfully adapted into film by Harold Pinter and directed by Karel Reisz.[6] In this film, the three possible endings were turned into parallel narration of Victorian romantic relationships and modern love struggles. This classical art film might not fulfill the desire of some of the audiences' expectation of interactive choices. The plot and ending of the film had already been decided in 1967, even before publishing the novel *The French Lieutenant's Women*.

The first interactive film *Kinoautomat*[7] was created by director Radúz Činčera,[8] whose narration relied on votes of the audience at the cinema. Based on a series of short documentary films in the Krátký film *Praha* (The Short Film of Prague) movie studio, Expo '67 in Montreal offered his chance in experimenting with the world's first interactive movie, which turned out to be his most famous work. In Czechoslovakia Pavilion at Expo '67 the audience became the exciting "players" in the interaction by voting. This was not only a forecast for the future of interactivity in motion graphics narration but also an indication of the new art form as a criticism to political systems. The logo of Expo '67 in Montreal could be seen as a metaphor of this first interactive film *Kinoautomat*, the structure not pointing only in one direction, while nine choices point to one final result. At nine points during the film the action stops, and a moderator appears on stage to ask the audience to choose between two scenes; following an audience vote, the chosen scene is played (see Figure 5). "The film is a black comedy, opening with a flash-forward to a scene in which Petr Novák's (played by by Czech actor Miroslav Horníček) apartment is in flames. No matter what choices are made by the audience at that point, the end result is the burning building, making the film – as Činčera intended – a satire of democracy".[9]

The design of voting in the first interactive film is very important. It not only connected the choices of the audience, it pushed the cinema toward a new narrative environment. Green and red represented the different choices, with the audience required to wave their physical boards colored green or red. We made a similar voting system in class at art school, Renmin University of China. Each time the color vote motivated the classroom towards more passionate communication, students will look at each other and figure out their own voting position in the group. This experience inspired one of the students' redesigns of the poster of this first interactive film (see Figure 6). Therefore, we can imagine the interesting atmosphere at the Czechoslovakia Pavilion during Expo '67. We could see common points between the novel *The French Lieutenant's Women* and with three

---

6   *The French Lieutenant's Women*. Directed by Karel Reisz, 1981.

7   *Kinoautoma*. Directed by Radúz Činčera, Ján Roháč and Vladimír Svitáček, 1967.

8   Radúz Činčera (17 June 1923, Brno–28 January 1999, Prague) was a Czech screenwriter and director.

9   Jeffrey Stanton "Experimental Multi-Screen Cinema", https://www.westland.net/expo67/map-docs/cinema.htm (accessed January 15, 2019).

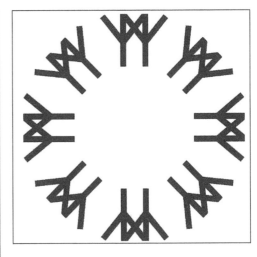

Figure 5: The logo of Expo '67 in Montreal could be seen as a metaphor of the interactive film *Kinoautomat*, the structure not pointing only in one direction, but nine choices point to one final result.

alternative endings and the interactive film *Kinoautomat* with nine choices: These works not only showed the shifting roles of authorities, in that the audience became the directors, but they also inspired the invention of a generative system to create art works automatically.

Generative Animation as Digital Evolution

Since 1980, the computer has been developed as a new device to automatically create artistic works and even new digital lives. In generative artistic animation artists transformed the experiments of narrative structures into digital evolution. From electronic abstracts (*Aquarelles*)[10] to tracing models of marine organisms (*Float*,[11] *Embryo*,[12] generative animation gave more freedom to discuss the issue of evolution. Tom De Witt created *Aquarelles* in 1980. During that time, De Witt set up Video Synthesis Laboratory with Vibeke Sorensen and Dean Winkler, realized a series of music video and abstract pieces using facilities at Teletronics, EUE-Video, and R.P.I.

10   *Aquarelles.* Directed by Tom DeWitt, Vibeke Sorensen and Dean Winkler, 1980.

11   *Float.* Directed by Yoichiro Kawaguchi, 1988.

12   *Embryo*, Directed by Yoichiro Kawaguchi, 1988.

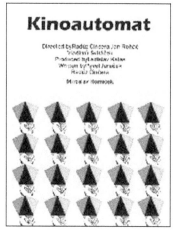

Figure 6: Wei KunYing (魏昆莹), student of the Art School Renmin, University of China, redesigned the film poster of the first interactive film *Kinoautomat*.

The meaning of *Aquarelles* was originally about painting using water color pencils, while in this abstract animation we can feel more about the colorful description of the deep ocean. Though there are no animal figures shown in this 7-minute length music video work, the stars, squares and circles seem more connected with mysterious marine organisms. This pioneering generative animation based on musical feeling might not be a coincidence, a very early example of a generative system could be traced back to Johann Philipp Kirnberger's *Musikalisches Würfelspiel* (Musical Dice Game) in 1757. Dice were used to select musical sequences from a numbered pool of previously composed phrases. This system provided a balance of order and disorder. Similarly, the colorful generative animations including *Aquarelles*, *Float* were set up to generate entire compositions with limited human intervention.

These colorful animation works were different from the abstract electronics in the first stages of digital art, which revealed the vivid lives made using the digital world. We would be a little shocked to see "too much" of a color palette in this animation, but it is understandable as in 1980, how excited digital artists would have been when they had so many vibrant choices! During the whole of the 1960s and 1970s most computer arts were about electronic abstracts in only black and white. This colorful abstract animation showed more feelings similar to game play, the artist himself enjoyed changing with the variable quantity setting in this program, the color and shaping must sometimes be beyond his own imagination.

Japanese animation artist Yoichiro Kawaguchi developed generated creatures that were digital representations of their real-world counterparts. In 1982, Yoichiro Kawaguchi created the first series of digital animation *Growth Model*[13] which was first exhibited at a SIGGRAPH conference. Additionally, Yoichiro Kawaguchi's digital animation works *Ecology II:*

13   *Growth Model*, Yoichiro Kawaguchi's organic animations based on the mathematical principles that underlie life. Exhibited in SIGGARPH in 1982.

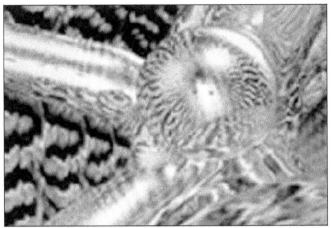

Figure 7: In 1987, Yoichiro Kawaguchi's awarded digital animation works
Ecology II: Float II offered the whole world a totally new 3D perspective of
digital creatures.

Float II[14] got a honorary mention at Prix Ars Elecronica 1987. Though only two years passed since the work of *Aquarelles* by DeWitt, Yoichiro Kawaguchi offered the whole world a totally new 3D perspective of digital creatures, with his very own style through a combination of organisms and revolution (see Figure 7).

There are some main differences between the digital world in *Aquarelles and* Yoichiro Kawaguchi's work, especially *Embryo* (1988), and these two works could be regarded as two stages of generative animation. Firstly, this depiction of digital lives transitioned from 2D to 3D, the details and the gestures made these fake lives more believable. Secondly, the theme of marine organisms was more apparent, and focused more on the variations of the combination of plants, animals, machines and sometimes features of human beings. Thirdly, the structure of the animated world was more connected with the cultural background religion of Buddhism, though it never shows the real figure of Buddha, the multiple layers of circles depict the feeling of reincarnation. And this kind of work is not more optimistic but seeps out a feeling of fear, being afraid of the mutated forms of the gene. Though it is also colorful, the color palette is not like the innocence of *Aquarelles*, it is more related to the Doomsday Complex and sexual culture from Japan.

After the example of works such as *Aquarelles* and *Embryo* in the 1980s, artist William Latham expressed himself by generating 3D computer models and organic life forms, which connected to both artistic expression and scientific thinking from the 1990s. Artist William Latham depicted the world of digital lives in a stricter way showing more persuasive details. He was inspired by Surrealistic painter Salvador Dali and Yves Tanguy, and created tangible textures of digital sculptures as a mixture of real creatures, organs

14   *Ecology II: Float II.* Directed by Yoichiro Kawaguchi, 1987, Honorary Mention 1987.

Figure 8: Artist William Latham's computer graphic work Mutation Raytraced on the Planet was exhibited at Prix Ars Electronica in 1992.

and dreams. He explored a system to develop his evolutionary drawings using digital lives (see Figure 8).

Unfortunately, the theory of generative animation was formed after methods currently associated with it were put into practice. This meant that artists were unable to follow a theory during the process of their creation. As a sub-field of generative art, the concept and analysis of generative animation benefited from the discussion of scholars and artists led by Philip Galanter. In 1993, Philip Galanter of New York University demonstrated four traits of generative art, one of the traits revealed the forces of animation in this world, which is a principle borrowed from the genetic system.[15]

1. Using randomization to create combinations or mixtures.

2. Using genetic systems to create forms of evolution.

3. A dynamic art developing with the change of time.

4. Based on the program running on the digital computer.

Randomization, genetic systems and time-based changes became the key-words of generative animation. This kind of theory exploration also pushed more artists, scientists and engineers together in conferences, exhibitions and global cooperation.

Over the past twenty years generated animation has developed into a variety of expanded forms. This paper distinguishes four categories as below:

A: E-Creatures Developed from Sculpture to 3D Printing

Yoichiro Kawaguchi was also a pioneer in developing digital creatures as colorful physical sculptures, which also expanded the look of digital art exhibitions. The creative process from animation model in virtual space into physical sculpture could be also reversed. Colorful and touchable sculpture

---

15 Philip Galanter, "What is Generative Art? Complexity theory as a context for art theory," *International Conference on Generative Art* (2003).

Figure 9: Generative animated works were also expanded into touchable installations including colorful physical sculptures by Yoichiro Kawaguchi, this is an example in Are Electronic Centre in 2009.

as a medium also expands the possibility of new ideas in generative animation (see Figure 9).

These large-scale creatures in sculpture form or even produced with 3D printing offered a different feeling to the audience. You feel you are so close to the life of a digital intellect and cannot contain the desire to touch or play with them. The huge gap between a human and a digital life in the blue-print of Yoichiro Kawaguchi was already forecast in a scene from the film *Arrival*[16] directed by Denis Villeneuve in 2016.

B: Virtual Immersive Generative Animation's Space

Some artists focused on how to combine generated animation with specific space from modern exhibitions to traditional architecture. *Fractal Flowers*[17] by French digital artist Miguel Chevalier is a new generation of virtual gardens. As described in the artist statement: "It's a new generator, which produces gigantic fractal flowers of different sizes, colors and shapes."[18]

In another Miguel Chevalier work *Dear World... Yours*[19] in 2015, he imagines a number of different graphic universes, which are generated in real time and use their own "digital" language to illustrate and interpret a wide variety of subjects including Academic Excellence, Health, Africa, Biology, Neurosciences, Physics, Biotechnologies. He collected the image data from

16   *Arrival*, directed by Denis Villeneuve, 2016.

17   *Fractal Flowers*, generative and interactive virtual-reality installation, Miguel Chevalier, since 2008, Source: http://www.miguel-chevalier.com/en/fractal-flowers-4?position=79&list=VXIgZIDoY06B wMNnGqRxfR3WpkW1PdMwA1WStQb854Y.

18   The artist statement of *Fractal Flowers* can be reached at https://vimeo.com/6947810 (accessed January 4, 2019).

19   *Dear World... Yours*, generative and interactive virtual-reality installation, Miguel Chevalier, 2015, http://www.miguel-chevalier.com/en/dear-world-yours-cambridge (accessed January 4, 2019).

the above fields and let the computer system make the decision to combine them as a projection on the splendid walls and ceiling of Cambridge church. In the architecture of the 16th century chapel through light, the randomly collected data turned into a generative interface, leading the audience into a magical and poetic atmosphere where science meets spirituality.

C: Connection between Space Transformation with Digital Lives

In 1987, Italian architecture designer Celestino Soddu explored generative art with the transformation of 3D models in the design project of Italian medieval cities. This project regarded architecture as an element of DNA, and developed its form according to their own liveness in a digital world[20].

D: Interactive Installation of Digital Survival

The topic of digital creatures presented in a series of exhibitions such as *Creative Machine*[21] in 2014 at London, it also brought the popular discussion to more audiences and participants. In this kind of event, the public got more chances to determine the species' survival created by computer. Artist Karl Sims regarded virtual reality as a discussion platform about evolution theory by Darwin, based on his years of experiments with the data-parallel particle rendering system and generative animation. *Genetic Images*[22] (1993) was firstly exhibited in Le Centre Pompidou in France. In the open frame, the audience was offered a variety of choices in the savior of the species on an arc of screens. *Galapagos*[23] (1995) stepped forward to challenge the evolution by digital installation. In *Galapagos*, exhibited in Tokyo Integrative Art Gallery, the players could choose creatures by their steps on one of 12 machines. When the audience stepped on one monitor, the other 11 monitors were blank, and this chosen monitor saved one species.[24] The individual choice brought a new dimension to the re-evolution in digital lives using interactive behavior. Each audience became a God, creating new lives in a digital world. But, when we look back at all these open-structure art works, there arises a question, how could the artists return the right of choices to the public? Are these decisions really made by the audience? Or are the audiences still in an even bigger control of the authority? Some people turned to digital games to find solutions, such as Will Wright and his creation *Spore*.[25]

---

20 This generative design project can be viewed at https://www.generativedesign.com/progetti/med.htm (accessed January 4, 2019). Creative Machine, http://doc.gold.ac.uk/creativemachine/artworks/ (accessed January 4, 2019).

21 On a generative modeling system in a computer, the designer was motivated by each slightly different presence.

22 *Genetic Images*, directed by Karl Sims, 1993, http://www.karlsims.com/genetic-images.html (accessed January 4, 2019).

23 *Galapagos*, interactive media installation, Karl Sims, 1993, http://www.karlsims.com/galapagos/ http://doc.gold.ac.uk/creativemachine/artworks/ (accessed January 4, 2019).

24 Thomas Dreher, "History of Computer Art." Chap. IV.3.2: Evolutionary Art of William Latham and Karl Sims, 2013, http://iasl.uni-muenchen.de/links/GCA_Indexe.html (accessed January 4, 2019).

Re-evolution as the Interface

The genetic system in a digital world is different from that of the real world, through interactive behavior with audiences, these digital creatures are experiencing an adventure of re-evolution. Re-evolution as a new creation connected generative animation and games, itself also became a unique interface offering unlimited choices. The game *Spore* could be regarded as a testament to re-evolution which was supported by the system complexity theory.

Will Wright attributes his success to the power of toys. He thought his gaming invention borrowed from the creativity of Maria Montessori School, putting games as toys, letting the players discover the answers and themselves in gaming. Will Wright also believes that games have the possibility to experiment using long-term issues in a short time, and this is the real power of games. He demonstrated this in the example of *Powers of Ten*[26] (1977) written and directed by Charles and Ray Eames, a film about how large the universe is and how powers of ten become so drastically distant. In only 10 minutes it presented the unbelievably mysterious and huge universe. The game *Spore* also hopes to explore the unknown universe in a short term and make the audience aware of the unique importance of their choices. Each choice by the player greatly affects the development of their own cultivated creatures' development, and each time a new game was started, the creatures were different from the previous ones, which gave a metaphor of parallel universes. In fact, the players construct the digital revolution together, each of their choices and creatures developing, determining whether the fate of their planet would be survival or invasion from other alien species. In some way, *"The game also accommodates more tangible preferences in that players may approach problems using diplomacy or violence as they see fit."*[27]

There is an interesting connection between the novel *The French Lieutenant's Woman*, and pioneer abstract animations such as *Aquarelles* and *Embryo* during the 1980s as well as the digital game *Spore* published in 2008: the ocean acts as a symbol of the mysterious and a source of new life. In the novel *The French Lieutenant's Woman*, Sarah was always standing on the shore looking towards the ocean, as shown by the author's reference of Hardy's poem "The Riddle" on the first page of the novel:

> "Stretching eyes west Over the sea, Wind foul or fair, Always stood she Prospect - impresses; Solely out there Did her gaze rest, Never elsewhere Seemed charm to be."[28]

25   *Spore*, Software published by EA Games, https://www.spore.com/ (accessed January 4, 2019).

26   *Powers of Ten*, directed by Charles and Ray Eames, 1977.

27   Robert Phelps is Chunning Guo's English teacher in Taiwan, as a player of game *Spore* he shared his experiences with this game in a conversation with Chunning in 2016.

28   Thomas Hardy, "The Riddle," quoted in John Fowles, *The French Lieutenant's Woman* (London, 1969), p. 3.

The attraction of Sarah is not only from herself, actually the whole image of Sarah standing beside the ocean became a riddle for Charles, and this is also a presentation of history itself like the deep ocean always dynamically shaped in the post-historical novel.

Aquarelles by Tom De Witt and *Embryo* by Yoichiro Kawaguchi echoed the diversity of the ocean from the more abstraction methodology. Though the latter provided audiences with more details and a 3D visual style, both of these works focused on the process of marine life growing and moving in the water. The whole process is also a symbol of circles in the bigger universe.

In the game *Spore*, Will Wright seems to have had bigger ambitions to also let the players themselves create an entire universe from a single-celled creature in the water (see Figure 10). Through vital abstract visual principles, *Spore* presented a process like evolution, which this paper would name as re-evolution, as this is a highly personalized, humorous and cooperative adventure. Users or players of this game would be encouraged to invest themselves in characters they created (with Wright's mind-boggling CG tools), and then with curiosity and passion to watch them encounter the fundamentals of life and nature. Wright would like to produce this entertainment content with a suspiciously educational function, with the contributions and choices of the players, this game develops as an original pedagogy to explore the possibility of re-evolution in digital systems.

The explorations about generative animation and re-evolution game are also inspired from cross-disciplinary research. In the book *A Legacy of Freedom*[29] by French geneticist Albert Jacquard, the author describes a unique species

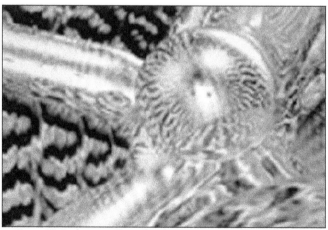

Figure 10: In the game *Spore*, the players can create an entire universe from a single-celled creature in the water.

29   Albert Jacquard, L'Héritage de la liberté De l'animalité à l'humanitude, Editions du Seuil, 1986. (Chinese translated edition), translated by Gong Huimin, GuangXi Normal University Press, 2005 July. p. 8.

of human beings who broke the barriers of the genome and their environment as they have the ability of free choice. The choices of Heterogeneity generated from the interface of re-evolution reflect the global scientific genetic experiments, which also raises questions related to scientific ethics. Will human beings who are nowadays occupied with using all kinds of interface in almost every moment become a new species: interface-being? These expanded animation explorations also raise questions related to scientific ethics with implications for decision-makers, particularly in the domain of genetic research and editing.

Bibliography

Boden, Margaret A. and Ernest A. Edmonds. "What is Generative Art?" *Digital Creativity* (March 2009), pp. 1–31.

Franke, Herbert W. "Mathematics As an Artistic-Generative Principle." *Leonardo* 22, no. 5 (1989), pp. 25–26, https://muse.jhu.edu/ (accessed January 4, 2019).

Galanter, Philip. "What is Generative Art? Complexity theory as a context for art theory." *International Conference on Generative Art* (2003).

Albert Jacquard, *L'Héritage de la liberté de l'animalité à l'humanitude*, Editions du Seuil, 1986. (Chinese translated edition), translated by Gong Huimin, GuangXi Normal University Press, 2005 July.

Ernst H. Gombrich. *The Story of Art*. 16, revised edition, London, 1995.

Hutcheon, Linda. *A Poetics of Postmodernism: History, Theory, Fiction*. New York, 1988.

Walker, John A. *Glossary of art, architecture, and design since 1945*. 3rd edition, London and Boston, 1992.

Wands, Bruce. *Art of the Digital Age*. London, 2006.

Part 4

Writing the Skies

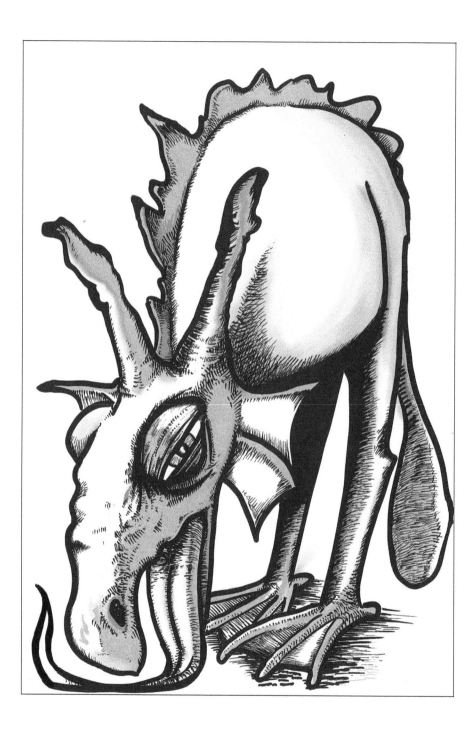

Chapter 7

Fell Beasts and fell beasts: The Making of a Monster in J.R.R. Tolkien's *The Lord of the Rings*

Damian O'Byrne

'Even of the sky above you must beware as you go on your way'
Elrond, *The Fellowship of the Ring*

Introduction: Concerning Fell Beasts

The skies of J.R.R. Tolkien's Middle-earth are bereft of the pegasi, griffins, hippogriffs and other beasts of the sky typically found in many sword and sorcery films. Perhaps this is because these creatures are all hybrids – the pegasus is a winged horse; the griffin half-lion, half-eagle; and the hippogriff a combination of horse and eagle. As such, any text containing such beasts would be instantly 'fantastical' – something Tolkien was keen to avoid.

Despite the absence of these iconic beasts of legend, the skies are far from empty during the Third Age of Middle-earth; it is simply that the flying creatures included in Tolkien's most famous works possess a grounding in reality that makes even the most fantastical of them seem somehow real. The armies of evil march beneath bats 'so dense that no light could be seen between their wings' (HOB, p.336); the armies of good are able to communicate with thrushes and ravens in *The Hobbit*; Saruman is 'spying out the lands' (LOTR, p.278) with flocks of large crows called Crebain; and the great eagles aid the forces of good numerous times throughout the stories.

Of course, the skies of Middle-earth are not entirely of our world and Tolkien does occasionally make a concession to the fantastical, perhaps most notably with Smaug, the mighty dragon who drives the narrative of *The Hobbit*. Whilst undoubtedly a great beast of myth and legend, Smaug is something of an outlier. The last great dragon remaining in Middle-earth, he is a reminder of a bygone age and, following his death during *The Hobbit*, no dragons remain to trouble the skies of Middle-earth during the events of *The Lord of the Rings*.

However, there remains another great beast of the sky in Tolkien's writing, at once perhaps the most recognisable and the most illusive – the Fell Beasts. Following the death of their horses, the Ringwraiths, or Nazgûl, mount great winged steeds, now most commonly referred to as Fell Beasts. I can vividly remember seeing a Fell Beast for the first time on the cover of *The Lord of the Rings Adventure Game*, a roleplaying game I was given when I was ten years old. The cover artwork was Angus McBride's painting *Éowyn, Merry & the Nazgûl* and I remember that the incredible image of Éowyn's iconic duel against the Witch-king atop his Fell Beast played a huge part in making me want to open up *The Lord of the Rings* and discover what other mysterious creatures could be found in Middle-earth. Countless subsequent hours spent poring over the pages of *The Lord of the Rings Adventure Game* ensured that McBride's painting was the formative image of the Fell Beasts for me, long before *The Lord of the Rings* was adapted for the big screen.

Nonetheless, even though many other Tolkien artists have helped to bring the Fell Beasts to life in countless paintings and illustrations over the past few decades, I would argue that it was not until Peter Jackson's three-film adaptation of *The Lord of the Rings* (2001–2003) that the Fell Beasts truly landed in the public consciousness. Much in the way that even casual fans know that *Star Wars* contains lightsabers or Indiana Jones wears a fedora and carries a whip, the Fell Beasts have become one of the most recognisable visual tropes of *The Lord of the Rings*, sitting alongside Gollum, Gandalf and the Shire as recognisably '*Rings*' even if you haven't seen the films. Jackson has long been a lover of creature features[1] and in the 2001 documentary *WETA Digital*, he revealed that 'one of the real motivations for me to want to make *The Lord of the Rings* was the monsters'. Whether it be the dramatic fight with the Cave Troll in *The Fellowship of the Ring* (2001), Sam's awe-inducing first sighting of the elephant-like Mûmakil in *The Two Towers* (2002) or the devastating attack of the Fell Beasts against Minas Tirith in *The Return of the King* (2003), Jackson's love of monsters is clear for all to see throughout his films.

Jackson's monster sequences are so visually spectacular that they are almost always included in trailers and adverts for the film trilogy and the Fell Beasts are no exception; it is now hard to imagine seeing any promotional sequence for the films which does not include at least a glimpse of one of these great beasts of the sky. Indeed, Jackson's films have made the Fell Beasts so utterly synonymous with *The Lord of the Rings* that it is very easy to forget that they actually play a much smaller, but arguably more effective, part in Tolkien's book. The Fell Beasts of Tolkien's novel are a fascinating part of his legendarium and their gradual introduction to the story is a masterclass in simmering narrative tension; their origins are shrouded in mystery and, perhaps most notably, the very etymology of the name Fell Beasts is a compelling journey through 70 years of Tolkien criticism and adaptation. This chapter will seek to explore the development of the Fell Beasts across a range of media, in a quest to discover the true nature of these mysterious creatures.

'What new terror is this?'
Legolas, *The Two Towers*

The Coming of the Fell Beasts to Screen and Page

Given that it was Peter Jackson's films that have made the Fell Beasts such an important part of the iconography of *The Lord of the Rings*, their on-screen depiction is an instructive place to begin. However, to truly understand the impact of Jackson's cinematic introduction of the Fell Beasts in *The Two Towers*, we must first briefly explore the story of the Ringwraiths in *The Fellowship of the Ring*. The Ringwraiths are the primary antagonists of the first half of *The Fellowship of the Ring*, hounding Frodo and his companions from horseback all the way from the Shire to Rivendell. However, despite Aragorn's ominous warning to Frodo that the Ringwraiths 'will never stop hunting you', they are defeated at the Ford of Bruinen and washed away down the river shortly after, at which point they disappear entirely from the narrative until they reappear in the next film in a *tour de force* of callback filmmaking.

The first hint we get of the return of the Ringwraiths comes via the soundtrack as Frodo, Sam and Gollum pass through the Dead Marshes in *The Two Towers*; as Frodo tries to learn more about Gollum's past, we suddenly hear the scream of the Nazgûl, a unique diegetic sound that instantly transports both the characters and audience back to the first half of *The Fellowship of the Ring*. This iconic sound effect is particularly important as the audience is being reminded of a creature that was last seen on screen nearly three hours earlier in the narrative. This instant recognition was even more important during the film's original cinematic release as it had been a full year since the Ringwraiths were seen on screen and it was vital to remind the audience of their impact on the narrative.

Shortly after we hear the scream, the film quickly introduces more cues to remind the viewer of the Ringwraiths. Sam yells, 'Black Riders!', a reminder that in *The Fellowship of the Ring* the Nazgûl were mounted on horses. Frodo clutches his shoulder where one of the Ringwraiths stabbed him in the previous film and Jackson even includes a five-shot flashback to the first film to remind us of the incident. This compelling combination of sound effects, dialogue, performance and editing serves to remind the audience of this terrible threat and prepare them for the return of the Black Riders. We then see a close-up shot of a pair of hands holding reins, followed by a close-up shot of the Ringwraith's head, confirming our suspicions. The Black Rider turns towards the camera and we see a reaction shot of Frodo and Sam. It is only when the shot comes back to the Ringwraith that the camera pulls out and we see that this Ringwraith is not mounted on a horse, but rather a huge flying creature, giving us our first glimpse of Jackson's vision of a Fell Beast.

Gollum voices the shocked feelings of the audience by crying 'Wraiths! Wraiths on wings!' before the Fell Beast flies into the distance. The Hobbits are safe for now, although the viewer can be sure that Chekhov's Fell Beast will be returning at some point in the narrative. So it is that in Jackson's adaptation the Fell Beasts are introduced with no preceding suspense. Sam and quite possibly the audience believe that the Ringwraiths are dead and that there is no longer anything to fear from them. The revelation that the Ringwraiths have survived their earlier defeat and are now mounted on huge flying creatures is both cinematically exciting and a great way for the digital effects artists to introduce a new monster into the narrative.

Whilst the thrilling and unexpected appearance of the Fell Beasts works incredibly well cinematically, the book takes a very different approach to their introduction. Indeed, before Tolkien fully reveals the Fell Beasts at the Battle of the Pelennor Fields, their introduction is teased far more gradually than in Jackson's films – a study in creeping dread. In the extended editions of the films, three hours of screen time passes between the Ringwraiths' defeat at the Ford of Bruinen and their reappearance at the Dead Marshes, roughly 25% of the story (the three extended editions have a combined runtime of 11.5 hours). Conversely, in Tolkien's book, after the Ringwraiths are defeated in the 'Flight to the Ford' chapter (p.192), it is some 400 pages until they are seen by Sam and Frodo in the Passage of the Marshes chapter (p.606), roughly 40% of the story later (the 1995 Harper-Collins edition has 1,008 pages). In other words, the length of time before their reappearance is significantly longer in the books than in the films. In addition, rather than the entirely unexpected, if visually spectacular, reveal of the Fell Beast flying over the Hobbits of the film, there is instead a creeping sense of unknown menace growing throughout the course of the 400 pages that the Ringwraiths are absent from the narrative.

The first hint of both the return of the Ringwraiths and the existence of the Fell Beasts can be found as the companions travel down the river Anduin:

> A dark shape, like a cloud and yet not a cloud, for it moved far more swiftly, came out of the blackness in the South, and sped towards the Company, blotting out all light as it approached. Soon it appeared as a great winged creature, blacker than the pits in the night (LOTR, p.378).

Gandalf later tells Legolas that the creature they saw was 'a Nazgûl, one of the Nine, who ride now upon winged steeds' (LOTR, p.487). Later in the story, after the main characters have separated into divergent storylines, we are given more glimpses of these mysterious beasts; as the companions are travelling towards Helm's Deep 'a vast winged shape passed over the moon like a black cloud. It wheeled and went north, flying at a speed greater than any wind of Middle-earth' (LOTR, p.581) and Dúnhere goes on to describe 'a flying darkness in the shape of a monstrous bird' (LOTR, p.776). Alongside these tantalising glimpses of the mysterious 'winged steeds' of the Nazgûl, we also hear the Orc Grishnákh ominously say that Sauron 'won't let [the winged Nazgûl] show themselves across the Great River yet, not too soon. They're for the War – and other purposes' (LOTR, p.442).

So it is that by the time we reach Sam and Frodo in The Dead Marshes in the book, we have already had at least four hints of this new enemy and there is an increasing sense of dread surrounding them. The Hobbits hear the same terrible scream as they do in the film which 'pierced them with cold blades of horror and despair, stopping heart and breath' (LOTR, p.593) and Sam suggests that 'like a Black Rider it sounded – but one up in the air, if they can fly' (LOTR, p.595) but unlike in the film, the creatures remain out of sight to both the Hobbits and the reader. Soon after, 'a vast shape winged and ominous...[passed] right above them, sweeping the fen-reek with its ghastly wings' (LOTR, p.615) and later still, Sam sees:

> A dark bird-like figure wheel into the circle of his sight, and hover, and then wheel away again. Two more followed, and then a fourth. They were very small to look at, yet he knew, somehow, that they were huge, with a vast stretch of pinion, flying at a great height (LOTR, p.630).

Over 250 pages have now passed since the rumour of these creatures first troubled the reader and the characters and, even as they slowly come into focus, the audience is still no closer to knowing their true nature. When the narrative then passes from Frodo and Sam to Pippin in the city of Minas Tirith, the dread of these winged creatures remains a constant, creeping threat. On the walls of Minas Tirith, Pippin recognises the scream of the Nazgûl from earlier in his adventure, much in the same way as Frodo and Sam, although it has now 'grown in power and hatred' (LOTR, p.791). As he looks for the source of the sound, he sees 'five birdlike forms, horrible as carrion-fowl yet greater than eagles' (LOTR, p.791) and Beregond describes them as 'fell things' (LOTR, p.791) – although notably not as Fell Beasts.

So it is that despite the fact that our heroes are now spread all over the realms of Rohan, Gondor and Mordor, the rumour of these terrifying winged creatures reaches all of them 'piercing the heart with a poisonous despair' (LOTR, p.791). Given how prevalent these creatures now are in today's visual depictions of *The Lord of the Rings*, it's very easy to underestimate what a powerful piece of storytelling this was on Tolkien's part. Despite being teased across the course of 400 pages, all the reader can so far ascertain is that they are fast-moving, dark, vast, monstrous bird-like creatures; without access to the countless visual depictions of the creatures to which we now have access, one can only imagine what manner of dark and terrifying creatures the mind of the reader would have conjured up during the first thirty years of publication.

The trope of withholding the reveal of a monster for as long as possible is now a common one in cinema; *Jaws* (1975), *Alien* (1979), *Jurassic Park* (1993) and many others have turned the art of not revealing the 'big bad' for as long as possible into an art form. Sometimes this decision is enforced by the production, such as how Spielberg was forced to make less of use of the shark in Jaws than he had planned due to the well-documented technical issues, but it has also become an increasingly common way for film-makers to build tension. Ahead of the release of *Godzilla* (2014), Bryan Cranston actually cited Jaws as an influence on his new film's approach to the monster:

> They're taking a very restrained approach to this, so much like Jaws did –
> Steven Spielberg didn't always show the beast, yet the essence is present and
> it's there and it's moving and you know and it's creepy – so the tension will
> mount for sure (Morawetz, 2014).

Of course, this cinematic technique dates back far further than the modern blockbuster, to the early days of horror cinema. Dewitt Bodeen, the screenwriter of cult horror *Cat People* (1942), said that 'that which [the audience] cannot see fills him with basic and understandable terror' (Bodeen, 1963, p.215) and, whilst cinema has a uniquely powerful ability to develop this sense of 'basic terror', particularly through its use of sound and shot selection, it is fascinating to see Tolkien using the same technique to such great effect in his novel.

Of course, Tolkien does ultimately reveal the Fell Beast in all its hideous glory as the Witch-king dives from the sky whilst riding one of the creatures during the Battle of the Pelennor Fields:

> It was a winged creature: if bird, then greater than all other birds, and it was
> naked, and neither quill nor feather did it bear, and its vast pinions were as
> webs of hide between horned fingers; and it stank. A creature of an older world
> maybe it was, whose kind, lingering in forgotten mountains cold beneath the
> Moon, outstayed their day, and in hideous eyrie bred this last untimely brood,
> apt to evil. And the Dark Lord took it, and nursed it with fell meats, until it
> grew beyond the measure of all other things that fly (LOTR, p.822).

And so we are finally presented with this vision of horror some 450 pages after that 'great winged creature' was first noticed by the Fellowship. Tolkien's dramatic reveal of his monstrous creature is saved for one of the most iconic moments in his story as Éowyn defends King Théoden from the assault of the Witch-king astride his monstrous steed. With the reader finally given a clear description of the Fell Beast, we can start to construct an image in our minds. It has 'neither quill nor feather', with a beak and claws and is large enough to bear a human rider, indeed it is larger than 'all other things that fly' meaning that it must be larger than the great eagles which frequently bear riders in Tolkien's works. However, whilst the reader can finally begin to conjure up a clear image of the beast, its true nature is still tantalisingly mysterious; the fascinating use of 'a creature of an older world maybe it was' implies that even the omnipotent narrator is dumb-founded by exactly what manner of creature the Witch-king is riding. However, whilst there is no further description of the Fell Beasts in the published version of *The Lord of the Rings*, we can gain further insight into the nature of these mysterious creatures by examining Tolkien's early drafts.

'They have come! Take courage and look!'
Beregond, *The Return of the King*

Forging the Beast with Pen and Paint

From 1983 to 1996, Christopher Tolkien published *The History of Middle-earth*, a 12-volume series that took an exhaustive look at the long develop-ment of his father's Legendarium, exploring early discarded drafts and offering fascinating glimpses of what might have been. Within volumes 7 and 8, *The Treason of Isengard* and *The War of the Ring*, we can find the earliest conceptions of the Fell Beasts which can offer an instructive insight into the kind of creature Tolkien was envisioning. The earliest mention of the creatures is in an early plot sketch from 1939, some 15 years before *The Two Towers* was eventually published and, at this earliest conception, Tolkien writes how 'the Black Riders have now taken form of demonic eagles and fly before host, or [ride] vulture birds as steeds' (TOI, p.208). Throughout *The Treason of Isengard*, Tolkien makes at least seven more references to the winged steeds as vultures and it is clear that at this point they are simply very large versions of the birds we are familiar with from our own world. At one point there was even a conceit that Sam would actually 'beat off' a vulture at the very climax of the story and this at least tells us that these winged steeds are small enough to be seen off by the attentions of one very angry Hobbit.

By the time Tolkien had actually begun to write the Passage of the Marshes sequence in 1944, the Fell Beasts are no longer commonly referred to as vultures. Instead, within *The War of the Ring* (2002), we find mentions of 'batlike shapes' (p.73), 'a great dark shadow like a huge bird' (p.107), 'the winged flier' (p.232), 'the dreadful wings' (p.329), 'winged steeds' (p.331), 'the great bird' (p.365), 'the huge vulture form' (p.365) etc. It would seem that Tolkien's conception of the nature of these beasts had changed over the course of the preceding five years and, despite the fact that Frodo describes them as 'great carrion birds' (LOTR, p.631) in the published text, Tolkien had clearly developed his conception of the creatures from large versions of our own vultures and towards a slightly more mysterious and fantastical creature, very much not of our world.

Perhaps as a result of Tolkien's final description of the creature as having 'neither quill nor feather, with a beak and claws…a creature of an older world maybe it was', for the first 50 years after the publication of *The Lord of the*

Rings, the most common interpretation of the Fell Beasts seems to have been of something resembling a pterodactyl; this is certainly how the Fell Beasts were most often depicted in artwork prior to Jackson's films, not least of all in Angus McBride's iconic painting on the cover of *The Lord of the Rings Adventure Game* that had such a profound impact on my childhood. However, whilst Tolkien himself said frustratingly little about the Fell Beasts after the publication of *The Lord of the Rings*, he did reveal that this interpretation was not precisely what he had in mind and, in response to the question of whether or not the Witch-king rode a pterodactyl, he replied in a 1958 letter that:

> I did not intend the steed of the Witch-King to be what is now called a 'pterodactyl', and often is drawn (with rather less shadowy evidence than lies behind many monsters of the new and fascinating semi-scientific mythology of the 'Prehistoric'). But obviously it is pterodactylic and owes much to the new mythology, and its description even provides a sort of way in which it could be a last survivor of older geological eras' (Carpenter, 2006, p.282).

There is a subtlety of language at work in Tolkien's response here; he is clear that the Fell Beast is not what we know as a pterodactyl, the informal term for any number of pterosaurs, the flying reptiles that patrolled the skies of the Mesozoic era. Instead, he is saying that the creature may well be similar in appearance and biology but is a creature exclusive to the history of Middle-earth. However, if we examine the work of those artists who depicted the Fell Beasts prior to the release of Jackson's films, we can see that there is a broad consensus that these creatures should indeed resemble pterosaurs. Angus McBride's *Éowyn and the Witch-king*, *Éowyn and the Nazgûl* by Ted Nasmith, the Fell Beast featured in Ralph Bakshi's 1978 animated film and a great number of other interpretations, all essentially depict the Fell Beasts as large pterosaurs. Whilst there have also been many illustrations of these creatures that have diverged from this interpretation, there is no doubt that the work of the most high-profile Tolkien artists will have helped to steer the court of public opinion towards the perception that the Fell Beasts should look very much like pterosaurs.

One notable exception to this interpretation can be found in the work of John Howe, perhaps one of the most influential Tolkien illustrators and certainly so over the past 35 years. Howe's paintings of the Fell Beasts, in particular *Éowyn and the Nazgûl* (1991), *The Dark Tower* (1989) and *Barad-Dûr* (1994) give the creatures an altogether different appearance. Gone is the long, pterosaur-like beak and in its place is a far more frightening and monstrous head, more akin to a cross between a Komodo dragon and a snake – the influence of this unique take on the appearance of the Fell Beasts cannot be underestimated. Howe's artwork rose to prominence in the official Tolkien Calendar, published annually since 1973 and described by Harper-Collins as 'an established publishing event, eagerly anticipated by Tolkien fans the world over' (2023). The calendars have had a huge influence on the visual perception of Tolkien's works and, in the 40 years that the calendar has been published, only Alan Lee and Ted Nasmith have had their work featured more times than Howe. The ubiquity of Howe's paintings in these calendars led Peter Jackson to draw on his work for the conceptual design of his film trilogy; indeed, in the foreword to *Myth and Magic: The Art of John Howe*, Jackson claimed that Howe's artwork was instrumental in bringing the films to the screen:

> Back in 1995, when Fran Walsh and myself were first thinking abut making a film adaptation of *The Lord of the Rings*, we were inspired by John Howe's paintings in the 1991 Tolkien Calendar... We used John's artwork in presentations, trying to interest Hollywood in the idea of a Tolkien movie and it worked because we got the money. Without his knowledge, he was already playing an important role in bringing *The Lord of the Rings* to life (Howe, 2001, pp.4–5).

Jackson's decision to draw so heavily on Howe's work in his film trilogy is a turning point in the development of the Fell Beasts. In the foreword cited above, Jackson also directly describes Howe's painting of *Éowyn and the Nazgûl* (featuring a Fell Beast) as looking like a still image from a film and it was Howe's interpretation of the Fell Beasts – which discarded the traditional pterosaur tropes – that became the driving design motif behind the Fell Beasts that would appear in Jackson's films. Indeed, if you look at Howe's *Éowyn and the Nazgûl* today, it looks like a piece of concept art for the film trilogy, not like a painting completed a decade earlier, so closely does it resemble the creatures shown on screen. Whilst there were obviously minor changes between Howe's original artwork and the Fell Beasts that would ultimately appear on the screen, his design influence is undeniable and the Fell Beasts that lead the assault on Minas Tirith in Jackson's *The Return of the King* are Howe's Fell Beasts.

And thus from Tolkien's early concept of the creatures as large vultures, through a multitude of depictions of them as pterosaurs and finally by way of John Howe's driving creative force – we can trace the visual development of the creatures that led to the design popularised by the films. However, whilst we may now have reached some sort of consensus on the appearance of the Fell Beasts, the entire study of these creatures can be turned on its head by acknowledging the fact that they are not actually called Fell Beasts...

'It is the sign of our fall, and the shadow of doom, a Fell Rider of the air'
Pippin, *The Return of the King*

To Fly or to Fly and to be Fell or be fell?

This claim might seem somewhat counterintuitive given that I have spent the first half of the chapter very deliberately using the term Fell Beast, but the simple fact is that Tolkien never refers to them by this name. Whilst

Tolkien does speak of 'the carcase of the fell beast' (LOTR, p.825) and how the bodies of those slain in the battle were lain 'apart from their foes and the fell beast' (LOTR, p.826), the lack of capitalisation here makes it clear that this is simply a description of a beast that was fell and not a proper noun. These are the only occasions in which Tolkien uses the terms fell and beast to describe the winged steeds of the Nazgûl and, tellingly, there is no entry for 'Fell Beast' in either the index of *The Lord of the Rings*, or the exhaustive 484-page index to the twelve-volumes of *The History of Middle-earth*.

In other words, rather than naming the winged steeds of the Nazgûl Fell Beasts, Tolkien is simply using the word fell to describe them. The SOED defines fell as 'Fierce, cruel, ruthless; terrible, destructive' and, in *The Lord of the Rings: A Reader's Companion*, Hammond and Scull suggest that Tolkien is using the word to mean 'Dreadful, terrible' (2005, 110). A brief internet search for the origins of the word suggest that it may come from the Old French word *fel* meaning 'cruel, fierce, vicious,' or from the Medieval Latin *fello* – 'villain' – so we can discern that Tolkien was simply trying to describe the nature of the Nazgûl's winged steed, not to name it.

Further evidence for this position can be found throughout *The Lord of the Rings* as Tolkien frequently uses fell as a descriptive term for a wide variety of other things; amongst many others, the Witch-king is described as a 'fell chieftain' (p.250), the Ringwraiths are described as 'fell riders' (p.747), Gimli describes the Paths of the Dead as having 'a fell name' (p.764) and Sauron is even said to have fed the winged steeds of the Nazgûl with 'fell meats' (p.822). Perhaps most tellingly, whilst Frodo has his vision on the Seat of Seeing upon Amon Hen, he sees the 'deadly strife of Elves and Men and fell beasts' (p.391) within the forest of Mirkwood; it is inconceivable that the fell beasts that Tolkien is referring to here are the winged steeds of the Nazgûl, in fact it is quite clear that he means the monstrous spiders who feature in *The Hobbit* – using the identical term to describe another evil creature.

In total, Tolkien uses fell in this context 43 times throughout *The Lord of the Rings* to describe all manner of evil places and creatures and yet no other character or creature has subsequently adopted the word as part of their name in the way that the winged steeds of the Nazgûl have. It seems as though Tolkien's two unguarded uses of the word 'fell' have resulted in a generation of fans extrapolating an entirely new name for one of his creations. Whilst this is a fascinating development in itself, it is not the first time that the fortunes of the beasts of Middle-earth's skies have been steered by a misunderstanding of Tolkien's use of language.

The question of whether or not Tolkien's Balrog has wings and whether or not those wings allowed it to fly has raged since the book was first published. The debate most often centres around the key lines in *The Lord of the Rings* that 'the shadow about [The Balrog] reached out like two vast wings' and 'its wings were spreads from wall to wall' (p.322). An oft-quoted piece of 'evidence' that they can fly is a line about Balrogs from the appendices of *The Lord of the Rings* that, 'Thus they roused from sleep a thing of terror that, *flying* from Thangorodrim, had lain hidden at the foundations of the earth since the coming of the Host of the West' (p.1046, emphasis added). However, rather than a description of the act of flight, this is far more likely to be colourful language to mean fleeing. Curiously enough, this is not the only debate concerning whether Tolkien's use of fly means to flee or to travel with wings.

In 2012, a Reddit user by the name of VulcanDeathGrip published a fan theory which suggests, at great length, that Gandalf had planned to use the eagles of Middle-earth to transport the Ring to Mordor, based entirely on his single line of dialogue, 'Fly you Fools'. This iconic line of dialogue represents Gandalf's last desperate piece of advice to the Fellowship before he falls to his death in the Mines of Moria. According to VulcanDeathGrip, this iconic line is not, as most people believe, a plea for the Fellowship to escape but is actually an attempt to divulge his secret plan to destroy the Ring by visiting the eagles. The original post is over 1,000 words long and goes into a lot of detail but, in summary of his own work, Vulcan says:

> Gandalf secretly planned on taking the Fellowship to where the eagles live and having the eagles fly them to Mordor. The eagles lived on the other side of the Misty Mountains but all the routes for crossing them were too dangerous and difficult, and Gandalf (along with his secret plan) ends up falling down a chasm in a battle with the Balrog. Just before falling with the Balrog he tries to surreptitiously tell them the secret plan but was too surreptitious and they didn't understand. When he came back as Gandalf the White he had forgotten many things, including the plan to meet the eagles (2012).

Whilst this is an enjoyable and diverting piece of fan fiction, it has no supporting evidence and is almost certainly not what Tolkien intended, as attested by the fact that Tolkien never mentioned it in any of his drafts or subsequent letters. In addition, 'fly' was a common British synonym for 'run' in the 1940s when *The Lord of the Rings* was written and, as Jim Allan has argued, Tolkien actually uses the word fly to mean 'run' at several other points throughout *The Lord of the Rings*.

In 2013, Allan conducted a conclusive investigation into Tolkien's use of the words flying and fleeing throughout the text of *The Lord of the Rings*, which, whilst primarily intended to challenge the 'evidence' that the Balrogs could fly, also serves as a convenient rebuttal of VulcanDeathGrip's fan theory. Allan discovers that, whilst 'the word flying is used to mean "moving through the air" 19 times... [it is also] used 20 times to mean either "fleeing" or "moving swiftly over the ground"' (2013, p.44). Here we can observe a situation where Tolkien's outdated usage of 'fly' to mean 'flee' causes confusion amongst Middle-earth fans on at least two occasions. I would argue that the same has happened with fell – the use of which to mean dreadful or terrible has largely fallen out of common parlance – and so, to today's readers, it might seem more likely that, rather than being referred to as a beast which is fell, Fell Beast is actually the name of the creature.

'They say that men cower with fear as it passes, men who have never feared any enemy before'
Aragorn, *The War of the Ring*

How the Movies Made a Monster

However, if the use of Fell Beast to describe the winged steeds of the Nazgûl were simply a case of some readers occasionally misunderstanding Tolkien's word use, it would not explain the increasingly commonplace use of the term. The simplest way to demonstrate the modern ubiquity of the name is to type Fell Beast into Google; the first page of results contains entries for Fell Beasts on *One Wiki to Rule Them All*, *Villains Wiki*, *Tolkien Gateway* and the *Middle-earth Encyclopaedia*; whilst the first of these four popular reference sites does point out that there is no canonical reason for these creatures to be called Fell Beasts, the other three do not and are happy to include this as the official name of the creatures. This, along with the countless other references to Fell Beasts that can be found via this simple internet search, suggests that Fell Beast is now a common and accepted name for the winged steeds of the Nazgûl but, given that Tolkien never used the term as a name, when and where did it originate?

There were just under 50 years between the publication of *The Return of the King* in 1955 and the release of Peter Jackson's film trilogy so there was plenty of time for the term Fell Beast to have emerged amongst fans and scholars of Tolkien before the release of the films. Whilst it may be impossible to prove that the term was not in use in colloquial language during this time, we can investigate whether it was ever used in Tolkien studies by examining the past issues of *Mallorn: The Journal of The Tolkien Society*. Between the first issue of *Mallorn*, published in 1970 and the release of Jackson's *The Fellowship of the Ring* in 2001, there were 39 issues of *Mallorn* published (and three issues of its predecessor *Belladonna's Broadsheet*). Published annually or biannually for over 50 years, *Mallorn* is a wonderful source of Tolkien criticism and analysis. In an attempt to discover how the winged steeds of the Nazgûl were referred to from 1970 to 2001, I searched these 42 issues for the following keywords: 'fell', 'beast', 'winged', 'steed', 'Nazgûl', 'Ringwraith' and 'Witch-king'. I also searched for the misspelled words 'Witch King' and 'Nazgul' in case of editorial oversight; the results were remarkable.

The first item of note was that, despite the fact that these creatures are a fascinating and mysterious part of Tolkien's world, there is almost nothing written about them at all. Indeed, in those first 31 years of *Mallorn*, the winged steeds of the Nazgûl are only mentioned nine times in total, and these are all passing mentions in broader articles with not a single submission being focused on the creatures. Of these nine mentions, there are two uses of 'winged steeds' (Lewis, 1989, p.33; Lewis, 1992, p.19) alongside single uses of 'Nazgul-mount' (Appleyard, 2001, p.40), 'winged creature' (Harvey, 1972, p.31), 'Nazgûl's Steed' (Scull, 1990, p.12), 'carrier of the Nazgûl' (Hickman, 1989, p.41), 'Nazgûl is later on the wing' (Ryan, 1987, p.17), 'flying beasts' (Askew, 1987, p.9) and 'pterodactyl steeds' (Dunsire, 1979, p.18).

There are two important observations that can be drawn from this analysis, the first being that there is no clear consensus on the name of these creatures. However, the second, and far more important, point is that, in the first 31 years of *Mallorn*'s publication, there is not a single instance of Fell Beast being used to describe these creatures. I then turned elsewhere in my quest to find any rumour of Fell Beasts but there seems to have been very little merchandise produced for these creatures prior to the release of Jackson's films so there are no collectibles which might offer a hint to any form of official or agreed name. However, the tabletop miniatures manufacturer Games Workshop did release a figure in 1986 which they called 'Nazgul on Winged Beast', again hinting at the absence of Fell Beasts in the lexicon of the time. Whilst further research could no doubt be conducted, I feel comfortable using the evidence above to proceed with my enquiry under the assumption that, prior to 2001, the term Fell Beast was not in common usage as a name for the winged steeds of the Nazgûl. Therefore, it would seem that the name must have originated from 2001 onwards with the release of Jackson's films but, on first inspection at least, this seems not to be the case.

The term Fell Beast is not used at any point in Jackson's *The Two Towers* or *The Return of the King*; in fact, the only description anyone gives of them is Gollum's cry of, 'Wraiths! Wraiths on wings!' which is faithfully transposed from book to film. The films don't even include Tolkien's two references to 'fell beast' (LOTR, pp.825–826) and so, at first glance, there is even less of a case for attributing the origins of the name Fell Beast to the films than to the book. This feeling is only corroborated by listening to the cast and crew audio commentary for *The Two Towers* during which Sean Astin (Samwise Gamgee) calls the creatures 'winged Ringwraiths', Dominic Monaghan (Meriadoc Brandybuck) describes them as 'half-bat half-elephant' and Peter Jackson himself calls them 'big winged beasts'.

Over a year later, in the commentary for *The Return of the King*, things have become no clearer and, if anything, they appear to have worsened with Jackson, Monaghan and Lawrence Makaore (The Witch-king) all referring to the creature as a Nazgûl – these comments are the most egregious given that the term Nazgûl categorically refers to the Ringwraiths and not to their steeds. The confusion of the correct name of the winged steeds of the Nazgûl amongst the cast is never better demonstrated than when Astin attempts to praise the work of the digital artists who brought the creatures to life by saying, 'This guy here on the.. on the... *that* is just stunning!' – with no agreed nomenclature to fall back on, it seems that *that* must suffice.

However, whilst there is clearly some confusion about the name of the creatures on the part of the cast and director, the same cannot be said of the rest of the film's crew. John Howe (conceptual designer), Tania Rodger (workshop manager), Christian Rivers (visual effects conceptual designer), Richard Taylor (workshop supervisor), Jim Rygiel (visual effects supervisor), Joe Letteri (visual effects supervisor), Randy Cook (animation designer and supervisor) and Alan Lee (conceptual designer) *do* all use the term Fell Beast to describe the creatures. Similarly, in the section about the winged steeds of the Nazgûl in *The Soundscapes of Middle-earth* documentary, David Farmer (sound designer), Ethan Van Der Ryn (supervising sound editor) and Mike Hopkins (supervising sound editor) all refer to the creatures as Fell Beasts. And so we can observe an interesting trend where, whilst the cast and the director can't seem to agree about what to call the creatures,[2] the creative team *can*. Therefore, I would suggest that the use of Fell Beast as the name of the winged steeds of the Nazgûl was actually popularised not by Tolkien, not by the script of the adaptation, but by the creative team behind the films.

This assumption is borne out by examining some of the behind-the-scenes books released to accompany the films. In *The Lord of the Rings: Official Movie Guide* (2001) John Howe writes that 'watching the first fell beasts – the winged creatures used as steeds by Sauron's servants – taking shape in the computers was amazing' (p.28). In *Myth and Magic: The Art of John Howe* (2001), Alan Lee credits Howe as the designer of the 'Fell beasts' (p.140). In *The Lord of the Rings: The Making of the Movie Trilogy* (2002) Bryan Sibley describes the flying steeds of the Ringwraiths as 'the Nazgul fell-beasts' (p.166). In *The Art of the Lord of the Rings*, a collection of conceptual artwork produced for the films released in 2004 (a few months after *The Return of the King*), the term 'fell beast' is used at least nine times by Jeremy Bennet (visual effects art director), Daniel Falconer (designer), Shaun Bolton (designer), Ben Wootten (designer) and Gino Acevedo (prosthetics supervisor). Daniel Falconer in particular relishes the fact that 'the fell beasts were only scantily described in the books, leaving us with quite a bit of latitude when coming up with designs' (p.158).

It seems that, whilst the designers may have enjoyed the creative flexibility they had to create the look of the creatures, they overlooked their most influential impact, that of formally naming them. The reasons for this decision are immediately apparent to anyone who has tried to write about them at length, myself included; having moved from using the term Fell Beast in the first half of this chapter to the altogether more laborious winged steeds of the Nazgûl in the second half, I yearn for the simplicity of Fell Beast even at the expense of fidelity to Tolkien. One can only imagine the vast number of meetings, emails and concept sketches that led to the creation of any single scene in Jackson's films and the need to have a clear, identifiable name for each creature is easy to understand. The fascinating part of this is that no one seems to have noticed that they were inadvertently naming the creatures and I can find no single reference to anyone who claims to have decided officially to name the creatures Fell Beasts for the production. Instead, it seems to have happened organically without anyone really noticing; this is particularly surprising in the case of Alan Lee and John Howe, neither of whom used the term in their artwork in the 15 years prior to the release of the films but, as shown above, both of whom now use it freely.

'Soon their terror will overshadow the last armies of our friends, cutting off the sun'
Gandalf, *The Two Towers*

Conclusion: Concerning fell beasts

Of course, the real litmus test comes from the legacy of this change: what evidence is there that the design team's decision to name these creatures Fell Beasts is what led to it becoming the widely accepted term for the creatures today? Given that the name Fell Beast is never actually used in the films, a cynical viewpoint might suggest that picking it up from the behind the scenes features or accompanying books is actually an extremely unlikely source of influence for the casual viewer.

To investigate the potential impact of the films on the use of the name, I first examined the issues of *Mallorn* that have been published since 2001. Whilst it would be extremely satisfying to report that the 23 issues published since the release of the films were full of references to Fell Beasts, unfortunately this is not the case. In issues 40 and 41, there are three mentions of the creatures, although they are not referred to as Fell Beasts; instead we hear of 'the winged Nazgul' (Pretorious, 2002, p.38), 'a flying Ringwraith' (Noad, 2003, p.35), and how 'valkyries are often thought to have flown through the air on steeds... like the Nazgul' (Petit, 2002, p.40). It should be stressed that these two issues of *Mallorn* were published in 2002 to 2003, the period after the first two films were released but before the emergence of the books and documentaries that popularised the use of Fell Beast, which may go some way to explaining the absence of the name.

Unfortunately, in the 21 issues that have been published since, there are only two more mentions of the creatures, once as 'the winged beast of the Nazgul' (Forest-Hill, 2010, p.9) and once as 'a wraith and his winged steed' (Beal, 2017, p.20). So, whilst there have been no mentions of Fell Beasts, there have once again been so few mentions of the creatures that it is hardly instructive. I must confess that when the latest issue of *Mallorn* arrived, just before the submission of this chapter, I eagerly searched for any mention of a Fell Beast in the hope of the perfect conclusion; alas, the winged steeds of the Nazgûl were overlooked by the ranks of Tolkien criticism yet again and there remain only two passing mentions of the creatures in the last 19 years of *Mallorn*, making it all but impossible to observe a notable change, should one exist.

However, while there may not have been a discernible shift in the publications of the Tolkien Society in the last twenty years, there *has* been a vast swathe of merchandise made available to collectors, all of which uses the term Fell Beast to describe the creatures; Toy Biz produced a Deluxe Poseable Fell Beast, Play Along produced a Pelennor Fields with Fell Beast play set, Games Workshop currently sell a Winged Nazgûl which their webstore describes as 'a Ringwraith mounted on a Fell Beast',[3] Funko produce a Witch King on Fellbeast and WETA Workshop, the company who designed the costumes, props and creatures for Jackson's films, currently have a Fell Beast Bust statue available for sale.

Not only that, but the term has started to bleed into some Tolkien criticism, in *The Making of Middle-earth: The Worlds of Tolkien and The Lord of the Rings*, Christopher Snyder refers to 'the Black Captain's winged fell beast' (2013, p.161) and three authors in the 2011 collection *Picturing Tolkien: Essays on Peter Jackson's The Lord of the Rings Film Trilogy* use the term Fell Beast when referring to the creatures (Thompson, p.28; Kisor, p.106, Bogstad, p.238).[4] When combined with the overwhelming results of the Fell Beast web searches explored earlier, it seems inarguable that Fell Beast has become the commonly accepted term for the winged steeds of the Nazgûl over the last 20 years in non-academic circles.

Conversely, the examples from more recent issues of *Mallorn* (Forest-Hill, 2010, Beal, 2017) demonstrate an unwillingness from the academic community to adopt the Fell Beast name in favour of the more laborious, yet more faithful, use of 'winged steeds of the Nazgûl'. Sharin Shroeder (2011) summarises a perceived scholarly fear that film adaptations can colonise the books they adapt, somehow irrevocably changing the way the story is received by the public and tainting the original work. The misnaming of the winged steeds of the Nazgûl as Fell Beasts could certainly be argued to be an example of this – and a particularly egregious one given that it doesn't seem to have been a conscious effort on the part of the filmmakers. It is perhaps not hard to imagine the frustration of a Tolkien fan who, upon seeing a product for sale labelled as a Fell Beast, is aware that not only is it a change from the text but that it is unintentional. Perhaps as a result of this, most serious Tolkien criticism seems keen to avoid adopting the term, no doubt in their attempts to respect Tolkien's wishes to keep his work free from the 'vulgarization' (Carpenter, 2006, p.257) of film adaptation.

Tolkien wrote that 'the visible presentation of the fantastic image tends to outrun the mind, even to overthrow it' (TAL, p.49) and it is likely that he would have strongly objected to the way that Jackson transformed his mysterious, maleficent creatures into Hollywood monsters. In the same way that the shark of *Jaws*, Xenomorph of *Alien* and dinosaurs of *Jurassic Park* lose something of their aura once they are finally revealed, Jackson's digital

Fell Beasts, as spectacular as they are, lack much of the malevolent terror that Tolkien develops as his winged steeds haunt those 450 pages of *The Lord of the Rings* from a distance.

When Peter Jackson said that 'one of the real motivations for me to want to make *The Lord of the Rings* was the monsters', he presumably meant iconic creatures like the Balrog, trolls and Mûmakil and he would undoubtedly be delighted to hear that he had actually created a monster of his own, transforming Tolkien's winged steed, a mysterious 'creature of an older world' into the entirely new Fell Beast of the films. Despite the indisputable effectiveness of Jackson's digital Fell Beasts, the winged steeds of the Nazgûl were arguably at their most elementally terrifying before they had ever been visualised. Perhaps Tolkien was right when he wrote that, 'In human art fantasy is a thing best left to words' (TAL, p.49) and that his winged steed is altogether more effective when it remains unseen and unknown, not a Fell Beast but a fell beast.

Endnotes

1. Jackson cites the original *King Kong* (1933) as one of his favourite films and would go on to remake it in 2005.

2. A special mention must be given to Billy Boyd (Peregrin Took) – the one cast member to refer to one of the creatures as a 'Fell Beast'.

3. Games Workshop rules writer Alessio Cavatore was calling them 'majestic fell beasts' (p.112) as early as August 2003, when Games Workshop introduced the creatures to their game in issue 284 of *White Dwarf* magazine

4. Of course, these all come from a collection of work specifically focusing on Jackson's film adaptations so it could be assumed that the authors held an above-average familiarity with the behind-the-scenes features of the films.

Bibliography

Abbreviations of Tolkien's Works
HOB – Tolkien, J.R.R. (2012). *The Hobbit*, London: HarperCollins.

IND – Tolkien, C. & Tolkien, J.R.R. (2002). *Index*, London: HarperCollins.

LOTR – Tolkien, J.R.R. (1995). *The Lord of the Rings*, London: HarperCollins.

TOI – Tolkien, C. & Tolkien, J.R.R. (2002). *The Treason of Isengard*, London: HarperCollins.

WOTR – Tolkien, C. & Tolkien, J.R.R. (2002). *The War of the Ring*, London: HarperCollins.

TAL – Tolkien, J.R.R. (2001). *Tree and Leaf*, London: HarperCollins.

Other Works
Allan, J. (2013), 'A Flying Balrog'. *Mallorn: The Journal of the Tolkien Society*, No. 54, pp.43–44.

Appleyard, A. (2001), 'Other alternative Middle-earths'. *Mallorn: The Journal of the Tolkien Society*, No. 38, pp.39–40.

Askew, M. (1987), 'Council of Elrond Planning Sub-Committee'. *Mallorn: The Journal of the Tolkien Society*, No. 24, pp.8–10.

Beal, J. (2017), 'Tolkien, Eucatastrophe, and the Rewriting of Medieval Legend'. *Mallorn: The Journal of the Tolkien Society*, No. 58, pp.17–20.

Bodeen, D. (1963), 'Val Lewton'. *Films in Review XIV*, No. 4, pp.210–25.

Bogstad, J.M. (2011), 'Concerning Horses: Establishing Cultural Settings from Tolkien to Jackson', in

J.M. Bogstad, & P.E. Kaveny, *Picturing Tolkien: Essays on Peter Jackson's The Lord of the Rings Film Trilogy*, North Carolina: McFarland & Company, Inc, pp.238–247.

Carpenter, H. & Tolkein, C. (eds) (2006), *The Letters of J.R.R. Tolkien*, London: HarperCollins.

Cavatore, A.(2003), 'The Ringbearer Speaks', *White Dwarf*, September, No .284.

Dunsire, B. (1979), 'The Specula of Middle-earth'. *Mallorn: The Journal of the Tolkien Society*, No. 13, pp.16–20.

Forest-Hill, L. (2010), 'The Truth about Elves'. *Mallorn: The Journal of the Tolkien Society*, No. 50, pp.8–9.

Hammond, W.G. and Scull, C. (2005), *The Lord of the Rings: A Reader's Companion*, London: Harper-Collins.

HarperCollins (2023), *Tolkien Calendar 2023*. Online at https://harpercollins.co.uk/products/tolkien-calendar-2023-jrr-tolkien?variant=39683664969806, accessed 13 January 2023.

Harvey, J.M. (1972), 'The Huntsmen of Fiction'. *Mallorn: The Journal of the Tolkien Society*, No. 6, pp.22–32.

Hickman, A. (1989), 'The Religious Ritual and Practise of the Elves of Middle-earth at the Time of the War of the Ring'. *Mallorn: The Journal of the Tolkien Society*, No. 26, pp.39–43.

Howe, J. (2001), *Myth and Magic: The Art of John Howe*, London. HarperCollins.

Kisor, Y. (2011), 'Making the Connection on Page and Screen in Tolkien's and Jackson's *The Lord of the Rings*', in J.M. Bogstad, & P.E. Kaveny, *Picturing Tolkien: Essays on Peter Jackson's The Lord of the Rings Film Trilogy*, North Carolina: McFarland & Company, Inc, pp.102–115.

Lewis, A. (1989), 'Splintered Darkness'. *Mallorn: The Journal of the Tolkien Society*, No. 26, pp.31–33.

Lewis, A. (1992), 'Sauron's Darkness'. *Mallorn: The Journal of the Tolkien Society*, No. 29, pp.17–20.

Middle-earth Encyclopaedia. Online at http://middle-earthencyclopedia.weebly.com/fell-beasts.html, accessed 13 January 2023.

Morawetz, K. (2014), ET Canada Gets Destructive On Set Of 'Godzilla'. Online at https://etcanada.com/news/11276/et-canada-gets-destructive-on-set-of-godzilla/, accessed 13 January 2023.

Noad, C.E. (2003), 'Reviews of The Film: The Lord of the Rings: The Two Towers'. *Mallorn: The Journal of the Tolkien Society*, No. 41, pp.34–35.

One Wiki to Rule Them All. Online at https://lotr.fandom.com/wiki/Fellbeast, accessed 13 January 2023.

Petit, E. (2002), 'J.R.R. Tolkien's use of an Old English charm'. *Mallorn: The Journal of the Tolkien Society*, No. 40, pp.39–44.

Pretorious, D. (2002), 'Binary issues and feminist issues in LOTR'. *Mallorn: The Journal of the Tolkien Society*, No. 40, pp.32–38.

Russell, G. (2004), *The Art of The Lord of the Rings*, New York: Houghton Mifflin.

Ryan, J.S. (1987), 'The Wild Hunt, Sir Orfeo and J.R.R. Tolkien'. *Mallorn: The Journal of the Tolkien Society*, No. 24, pp.16–17.

Scull, C. (1990), 'On Reading and Re-reading *The Lord of the Rings*'. *Mallorn: The Journal of the Tolkien Society*, No. 27, pp.11–14.

Shroeder, S. (2011), '"It's Alive!": Tolkien's Monster on the Screen', in J.M. Bogstad, & P.E. Kaveny, *Picturing Tolkien: Essays on Peter Jackson's The Lord of the Rings Film Trilogy*, North Carolina: McFarland & Company, Inc, pp.116–138.

Sibley, B. (2001), *The Lord of the Rings: Official Movie Guide*, London. HarperCollins.

Sibley, B. (2002), *The Lord of the Rings: The Making of the Movie Trilogy*, London: HarperCollins.

Snyder, C. (2013), *The Making of Middle-earth: The Worlds of Tolkien and The Lord of the Rings*, New York: Union Square.

Thompson, K. (2011), 'Gollum Talks to Himself: Problems and Solutions in Peter Jackson's Film Adaptation of *The Lord of the Rings*', J.M. Bogstad, & P.E. Kaveny, *Picturing Tolkien: Essays on Peter Jackson's The Lord of the Rings Film Trilogy*, North Carolina: McFarland & Company, Inc, pp.25–45.

Tolkien Gateway. Online at https://tolkiengateway.net/wiki/Fell_beasts, accessed 13 January 2023.

Villains Wiki. Online at https://villains.fandom.com/wiki/Fellbeasts, accessed 13 January 2023.

VulcanDeathGrip (2012), Lord of the Rings: A theory about the eagle 'plot hole'. (Very long, it even has maps). Online at https://np.reddit.com/r/FanTheories/comments/130it2/lord_of_the_rings_a_theory_about_the_eagle_plot/, accessed 13 January 2023.

Filmography

The Lord of the Rings: The Fellowship of the Ring (2001) [Film]; Dir: Peter Jackson, USA: New Line Cinema, Wingnut Films.

The Lord of the Rings: The Return of the King (2002) [Film]; Dir: Peter Jackson, USA: New Line Cinema, Wingnut Films.

The Lord of the Rings: The Two Towers (2002) [Film]; Dir: Peter Jackson, USA: New Line Cinema, Wingnut Films.

The Soundscapes of Middle-earth (2002) [Film]; Dir: Michael Pellerin, USA: New Line Cinema, Wingnut Films.

WETA Digital (2001) [Film]; Dir: Michael Pellerin, USA: New Line Cinema, Wingnut Films.

Chapter 8

Buffalo Wings of Desire

Jay Russell

'God, I hate flying!'

I looked up from the craft-ale backwash I'd been quietly nursing – a peculiar Japanese brew called Shobijin that comes in tiny twin bottles. Truth be told, I'm more of a Bud man, but this didn't feel like a Bud kind of place. The guy who'd announced himself as a latter-day Erica Jong looked at me expectantly. (And I say 'guy' but in fact 'he' had no genitals at all – we'll come back to that.) He sipped something pale blue out of a delicate, horn-of-plenty glass, and flicked his lizard-like tongue across pink, chicken lips.

'You're kidding, right?' I said.

'Do I *look* like a kidder, friend?'

Fair dinkum to him, he most definitely did not come across as a missing Marx Brother. It wasn't just his crotchless crotch which suggested a certain sombre quality; it wasn't even the flaming red eyes, or the four black nipples, or the ornate, obsidian dagger with which he – rather messily – speared at his nachos.

It was the wings.

Folded they were, at the moment, but white as a pail of teat-fresh milk. The wings sprouted from the steel-cable musculature of his bare back – he was completely naked – arching from a point two feet above his massive, mis-shapen head right down that bronzed spine all the way to his cloven feet. The wings were downier than Robert Junior, each fluffy feather marked with an eye, and every one of them studying me with the earnest seriousness of a medieval monk on manuscript duty.

Correction: he wasn't *entirely* bare-ass: he wore a pair of custom, lime-green Crocs festooned with Disney Princess Jibbitz. (Jasmine proffered a particu-larly come-hither look, but then she always does.)

'So...*what*? Is that just bling?' I said, gesturing at his wings.

He looked askance at me. I've never been entirely sure what askance actually means, but every eye on those goddamn wings seriously askanced me, I was certain.

'I am Astaroth,' he recited, voice now booming. 'Brother to Amdusias and Agares. Father to Formeus and Fenex. Slayer of Belex. Castrator of Purson...'

'Ouchies!' I said. It passed him right by.

'...Master of Serpents, and Ruler of the Sixth Circle. I singe the heavens with my wings. I scorch the earth with my hooves. With *unholy* fury does my gaze condemn the unjust, the wicked, the impure.'

'So, you're, like, an Instagram influencer.'

'I am Astaroth Eternal, Eater of Souls. I am...'

He drew himself up to his full seven and a half feet.

'I. Am. A Duke. Of Hell.'

'Huh!' I said and narrowed my eyes at him. He continued to glare at me. 'Astaroth, was it?'

He gave me the slightest of nods.

'A-S-T-A-R-O-T-H?'

He gave me an annoyed, 'duh' look.

'Now, I could be wrong here, so do correct me, but don't I remember your name from *Bedknobs and Broomsticks*?'

Those broad shoulders slumped, his wings ruffling with a sound like baby farts. I'm pretty sure a dozen or so of those countless eyes rolled heavenward.

'*God*, I hate that movie,' he croaked. 'Stupid cartoon lion. Stupid cute little kids.'

'Angela Lansbury wasn't bad,' I said.

'Well, she's always good, isn't she,' he agreed.

Who could argue?

'So, what you drinking there, Asti?'

He looked into his empty glass. 'Curaçao and champers,' he muttered. I'd mutter it, too. Still, I signalled to the bartender for another round, tossing a Charon's obol on the counter in payment.

I stuck out my hand. Generally, I'm neither unjust nor wicked.

As for impure...

'I'm Marty,' I told him. 'Marty Burns.'

I needed a tinkle, so I left the Duke of Earl – or Hell, or whatever he claimed to be – propping up the bar. Following a windbag explanation of why Dante was a twat, he'd finally stopped yakking at me to hit on some three-breasted, more-cleavage-than-you-can-shake-a-stick-at Harpy perched on a neighbouring stool. I took my package with me – it was the whole reason for coming here, after all, and I wasn't about to leave it lying around for any Tom, Dick or Astaroth to snatch. I squeezed my way through the barflies, looking for the toilets.

I got halfway across the floor when I saw it.

We immediately locked eyes.

It had been a lot of years since I'd encountered one of them and it hadn't been pretty. It was way back in the depths of my failing PI days – the *Celestial Dogs* caper – when I had first learned that the universe is a bigger, badder and just plain weirder place than even a former child star could dream. One if its kind had damn near killed me back then. Not something you readily forget.

I stood face-to-face with a *tengu*.

Taller even than old Astaroth, this fiercest of Japanese demons loomed over me like an impending tax audit. Its yellow eyes widened upon seeing me and with a Noddy Holder shriek, it unfurled its thick, leathery wings. The tengu's long needle-nose dripped globules of turd-coloured snot which it hungrily lapped at with its two tongues. This tengu was the colour of spilled blood congealed in the summer sun. Its foul mouth gaped open, revealing curved lines of tiny scimitar teeth. Its lips parted in a perverse, Benicio del Toro smile.

I tensed.

Then I glanced down.

Pinned to the Tengu's pustulant chest I saw a yellow lenticular smiley badge. It was emblazoned with a bold hashtag: **#NotAllDemons**.

I offered a half-smile of my own and raised one hand in a gentle, beauty pageant contestant wave.

It waved back.

I really needed to whiz now, and hurried my way through a Wyvern hen party to continue looking for the little creature's room.

The process turned out to be more complicated than anticipated. Not the peeing bit – hell, I've been doing that on my own since I'm thirteen – but the choosing-which-toilet bit.

There were five different doors. Each with an equally indecipherable icon.

One had a kinda fish-like thing, with long tentacles emerging from the heads. I pressed my ear against the door and heard what sounded like Idina Menzel singing '*R'lyeh wgah'nagl fhtagn*'.

I passed.

Another door featured a flaming skull with a bloody eye-stalk trailing out of its mouth. Chunky, greyish-red liquid spilled out onto the floor from underneath, so I figured maybe that wasn't the way to go either.

Two others had symbols so alien and unsettling that I considered trying to simply absorb the urine back into my body somehow, but I can't even do Pilates, so...

The last door just had a blob on it – I think maybe the plaque had fallen off. Praying that *The* Blob wouldn't be inside, I cautiously gave it a push.

All good.

Clean tile floor. Gleaming sinks. A row of toilet stalls – okay, one of them featured three seats in a deeply puzzling array, like some Duchamp nightmare – but Bingo! A line of urinals along the wall.

There was no one else in the room.

I strode toward the far stall – bashful bladder, don't you know – and noticed that pictures of different faces had been cruelly plastered to the bottom of each urinal, targets for...golden opprobrium. It took me a minute but I recognized the first as JRR Tolkien.

What the hell?

I really shook my head at the second photo: Peter, Paul and Mary.

Whose john was this, anyway?

I was completely stumped by the third face, when I caught a whiff of something.

Sulphur?

The heavy toilet door flew open and I whirled around. Squeezing through the door frame – smoke rings billowing out of fist-sized holes in its wet snout – was...well...

A dragon.

Not a big one, mind, but it has to be said that even relatively small dragons make a hell of an entrance. This one was fire gold, with glistening flares of magenta and turquoise lighting up its scales. Its wings were tightly tucked into its heaving torso, and its sleek tail – with a barbed, red arrow-head at the very end – flicked back and forth like a happy chihuahua.

''Make a hole, Mac,' it said in a surprising Brooklyn accent, 'gotta empty the tank, toot-quick.'

It pushed past me and manoeuvred itself in front of that third urinal. I heard a little zip – what the...?!? – then there was a pause in which it looked over its shoulder at me.

'Après moi, le deluge, buddy. You've been warned.'

Then it let loose.

A couple of years back I was driving home to Hermosa Beach from Tarzana, up in the Valley. It had been pouring buckets all day and stupidly I took one of the low canyon roads. Out of nowhere this little trickle started spilling down from the hills, and before I could say 'My Prius only has two more payments' a flash flood enveloped the car. I barely got out the sunroof in time.

All I can say about the dragon's micturition is: 'My Prius only has two more payments.'

It wasn't a bad dragon, really. Hell, it thoughtfully carried its own supply of antibacterial wet wipes, and we needed almost all of them for the clean-up. It even gave me a 'sorry' gift: a small malachite gem – 'from my hoard,' it swore – but I'm pretty sure it was QVC.

'We good?' the dragon said.

I nodded and it started to cram its way back through the door.

'Hey!' I called.

It paused, raising an eyebrow.

'The pissoir you chose. The photo at the bottom?' I raised my hands in a questioning gesture.

"George RR Martin,' it hissed. 'Dragon's daughter my fucking ass.'

And off it went.

I completed my now desperate business in one of the stalls, but I didn't feel all that relieved.

★ ★ ★

The crowd had thinned out some – I'd vaguely clocked the announcement of several departures over the PA – but I still hadn't spotted my quarry.

Then I saw a sign I hadn't previously noticed, pointing to a room off the main lounge and three steps down: The Labyrinth.

I wandered in.

The snug was bathed in a dull glow, and Harry Chapin's 'Taxi' played softly over cheap speakers. A large window was set into the far wall, but the blind across it was louvered shut. Three round tables with Naugahyde banquettes occupied much of the floor space, bathed in red spotlights. One of the bulbs had blown, and sprawled face down across a table in shadow – littered with cans of Red Bull and a mostly empty fifth of speed-rack vodka – was a middle-aged fellow with golden blonde hair. It was greasy but still full and thick. As I stepped down into the room, a shot glass crunched underfoot, startling something tiny and quick which fluttered out of his matted hair. No bigger than a firefly, the little winged lady buzzed right past me, leaving a sprinkle of iridescent rainbow residue on my shoulder.

Fairy barf?

Pixie poop?

I don't know, but it sure did sparkle.

The man raised himself up. He looked at me, or through me – he had that thousand-yard stare. He scratched his head briskly – I sure hoped nothing else was living in his hair – and offered a gaping yawn. I could smell his breath ten feet away, but his perfect teeth were moonlight white. He wore a tatty tee shirt emblazoned with the words: IT'S THE SUN WOT WON IT.

After a minute he seemed to realise I stood in his line of vision and he rack-focused his gaze.

He blinked a couple of times.

I cleared my throat. 'I'm looking for Daedelus.'

'No one here by that name,' he said. 'No one at all.'

He had a voice like a tinkling, crystal Christmas tree ornament.

'I was told…'

'Don't know, don't care,' he interrupted again.

'I was given a description. Surely, you're…'

'I'm just a guy,' he said.

'Just a guy?'

'Just a guy.'

I sighed. 'Are you sure…'

'Call me…Steve,' he said.

'Your name is Steve.'

He nodded.

'Your name is Stephen Daedel…'

'Just Steve,' he said.

'Oh. Okay, Just Steve.' I pointed at the banquette and looked at him. He shrugged, so I went ahead and sat down. 'I'm Marty.'

'Everybody's somebody.' He picked up a dirty glass, poured himself two fat fingers of hootch and added a splash of flat Red Bull. He didn't offer to share. 'Everybody's somebody, and nobody cares,' he moaned.

'Oooh, Red Bull and self-pity. It's no curaçao and champagne, but it'll do. Can you get that with a woe-is-me chaser?'

'Heeeey, you're a real funny guy. I knew a real funny guy once.'

'Oh, yeah?'

'Yeah: I threw him face first off a tall tower. And he was my sister's kid. You I don't know at all.'

The background music changed: Tom Petty singing 'Free Fallin''.

'I'm Marty,' I said again.

He tossed the drink down his throat, eyeing me properly now. I returned the study. Even though he was seated, I could tell that he wasn't big, though he looked solid as granite. Cheap vodka – a lot of it, I reckoned – had mottled his olive skin, dulled his terrapin-green eyes. But a visible and slightly fearsome intelligence remained evident behind the 70-proof glaze. It was somewhat undercut by a tell-tale barbecue sauce apostrophe on his chin – the floor around him was littered with the gnawed bones of all-you-can-eat Buffalo hot wings – but it provided a useful reminder of who I was dealing with.

He surrendered the stare-down first, grabbing another Red Bull and ripping the top off the can with a practiced flick of his wrist. Using thumbs and forefingers, he began tearing it into thin aluminium strips, twisting the strips together in a blur. His hands were like two hummingbirds – I couldn't follow their motion. Then he held one hand out in front of me – palm up – the tiny sculpture he'd fashioned wobbling ever so slightly with his DTs.

A perfect pair of wings.

'Beautiful,' I whispered.

He snapped his fist closed around them. When he opened it back up there was nothing in his hand but a mash of cheap metal and a thin line of blood.

'The things I could do with these hands,' he said. But I don't think he was talking to me.

'Listen: I've brought you something,' I told him.

I picked up the package I'd been carrying around all this time. That I had agreed to bring to this place – this most unusual eyrie – in search of…Steve.

'It's a gift.'

'Ain't Christmas and it ain't my birthday.'

'It's from your son.' I placed the item on the table.

His manufactured front crumbled and he immediately reached for the vodka. Then he froze, that talented, shaking hand hovering just over the bottle. He lowered it slowly to the table and instead he very gently picked up the small box I'd brought for him.

'My son.'

'Yes,' I said.

'It's from my son.'

I nodded. He shook his head.

'My son…is dead.'

'Your other son.'

'My other…' He stared at the box in his hand. He blinked once and said, 'Is he…well? My other son.'

'Jim Dandy,' I told him. His brow furrowed. 'He's fine,' I added.

He nodded. 'After …after his brother…after the fall, I kept away. So, so long. I've been…here.'

Criminy, I thought: that's a lot of Red Bull and chicken wings.

'To lose one child is unbearable. To lose two...'

'He remembers you. He still thinks about you. Every day. He told me all about you.'

He gave another little shake of the head. 'How...?'

'I was in an accident. A manticore, a '67 Mustang, some magic mushrooms...it doesn't matter. But I met him – your son – at Cedars. He looked after me. Healed me. It's what he does after all.'

'Cedars? You met him in the trees?'

'Sorry. Cedars-Sinai. It's a big hospital, a ...place of healing. In LA."

'El. Ay?'

I let out a long exhalation. 'LA is...it's...it's the City of Angels,' I said.

He nodded again, but I'm not sure he understood. Hey, LA – who the fuck does? He traced the carving on top of the little box with his thumb. It was a caduceus: Winged Hermes' staff.

'Aren't you going to open it?' I asked.

The great craftsman's hands trembled as he stared at it. Sinatra started singing 'Come Fly With Me' in the background.

Slowly, he opened the box.

I couldn't see what was inside, but his eyes went wide and the corners of his lips turned up ever so slightly. He flicked a glance my way.

A light began to glow around him.

He held the box in his left hand and with his right he plucked out its gift.

A feather.

It might just have been the cheap downlighters, but the feather burned neon red. He held it ever so delicately by the very sharp tip of its calamus. He turned it slowly round, his eyes growing wider with every twirl. He stood up and held the feather above his head, a look of wonder washing across his suddenly youthful face.

'Bennu,' he whispered. 'Oh, bird of rebirth.'

'Nifty,' I said. Ever the poet.

'Thank you, my friend.' His face had transformed entirely. He looked younger, stronger...restored.

Before I could make with a snappy comeback – or even just 'you're welcome' – he dropped his arms to his sides. The feather continued to float in the air above him, still spinning.

It didn't fall. He stared at it in joy.

Then it exploded into flame. The flare was so bright, I had to turn away and so can't be sure, but I had a vision of fire darting like a streak of crimson lightning at Daedelus, striking him squarely between the eyes.

He yelled. Or laughed – I'm still not sure. Then he ran.

Knocking over the vodka and drink cans, he headed straight for the big window. He made a gesture and the blind flew away. Still running, a curtain of red igniting behind him like a comet's tail, he used the last table as a springboard, and with a mighty roar and leap, he burst through the glass and out into the void.

'Shit, shit, shit,' I said, running over to look.

The view was sublime.

Looking up: the inky black ocean of the cosmos lapping against the banks of Earth's atmosphere, stars dancing like so many flecks of foam.

Looking down: our own tiny star sinking like an ice cube into a single-malt horizon, golden sunset bathing me in a cascade of orange-red warmth.

I leaned out as far as I dared, looking for some tell-tale speck of Daedelus falling.

Nothing.

Nothing could be discerned in the endless darkness above, either.

Then I heard a laugh. Or something like it.

Just along that hairline crack between the light of this world and the dark of the infinite, I thought I saw something flying, soaring.

Born aloft by perfect wings.

Or maybe it was just fairy dust in my eye.

Then it was gone.

'Well…hell,' I said out loud.

I admired the view for another moment – no mere mortal could endure such splendour for any longer. I gingerly stepped down from the window to the filthy barroom floor.

Picking my way around the hot wings, without a glance back – never look back – I wandered out into the main lounge.

Everyone was gone.

Well, almost everyone.

Sitting at the bar was the tengu. It glanced over at me and slapped a vacant barstool with a taloned claw.

It put a gouge in the wood.

I sighed and walked over to the demon.

The son of a bitch was drinking a Bud.

The tengu tilted its head in a maybe-you'd-like-one-too gesture.

I nodded. I sat down, and it beckoned to the bartender.

'You know,' I said, pouring my beer. 'This could be the beginning of a beautiful friendship.'

Biographies

Illustrators

Lee Brooks has crafted each of the conference posters and book covers for *Beasts of the Deep*, *Beasts of the Forest* and the current volume *Beasts of the Sky*. He is a senior lecturer at St Mary's University, where he teaches on and convenes a wide range of practical and theoretical modules that combine his professional and academic experience. He has published 'Talent Borrows, Genius Steals: Morrissey and the art of Appropriation' in *Morrissey: Fandom, Representations and Identities* (Intellect Books) and has a second piece on the bequiffed bard of Manchester: Ambitious Outsiders: Morrissey, Fandom and Iconography upcoming in *Subcultures, Popular Music and Social Change* (Cambridge Scholars) and also has research interests in Disney, theme parks and animation.

Rupert Norfolk is a London-based illustrator that has been providing the images across all three books in the *Beasts* project. His contributions have provided rich imaginings of the many topics that have been covered. When not hunched over his drawing desk, you can find him haunting one of the many dark, dank London alternative music venues. He is a fully certified goth, and has an enviable record collection.

Contributors

Dr Francis M. Agnoli is an animation scholar with a special focus on depictions of racial identities. Previous publications include *Race and the Animated Bodyscape: Constructing and Ascribing Identity in Avatar and Korra* (University of Mississippi Press, 2023) and the edited collection *The Avatar Television Franchise: Storytelling, Identity, Trauma and Fandom* (Bloomsbury, 2023). His work has also been published in the journals *Animation: An Interdisciplinary Journal* and *Animation Studies* as well as in the edited collection *Fantasy/Animation: Connections Between Media, Mediums and Genres* (Routledge, 2018). Agnoli received his PhD in Film, Television and Media Studies from the University of East Anglia, and he is currently an adjunct instructor at Kirkwood Community College, where he teaches courses on film history and popular culture.

Dr. Chunning (Maggie) Guo
20 years of teaching and research in the field of New Media Art and Animation. She was a visiting scholar in Vancouver Film School, a resident researcher at Central Saint Martins of the University of the Arts, as well as a resident artist at Centre Intermondes of La Rochelle in France. From 2017 to 2018, she was selected by the Sino-Dutch scholarship as a visiting scholar in Radboud University in the Netherlands. From 2023, she was invited as a visiting professor in Geneva University in the Switzerland. She earned her PhD in independent animation. She was invited to present 30 of her papers in more than 15 countries, including at Animafest Scanner, SAS, IAC, Under the Radar, Expanded Animation, etc. Her collaborative animation work *Ketchup* has been screened at more than 20 international film and animation festivals and received the Best Short at the Chinese Independent Film Festival in 2014 and the NETPAC Award at the Busan International Short Film Festival in 2015. She has published more than 30 papers in local and international journals, including *Contemporary Cinema, Contemporary Animation, Aesthetics, Art Education, National Art, Croatian Cinema, Cartoon and Animation Studies, Epistémè*, and *The Global Animation Theory* published by Bloomsburg in New York.

James Keyes is member of the Irish Association for American Studies and the European Popular Culture Association, James Keyes has served as Senior Tutor in Film and Media Studies at University College Dublin. He has published and presented on a number of topics, including American silent and science-fiction cinema, the work of Irish filmmaker Lenny Abrahamson, and representations of the Bosnian War in film.

Sara Khalili is an independent animation filmmaker, storyteller, and poet. She holds a Bachelor of Arts in Visual Communications from the University of Tehran and a Master's degree in Animation from Tehran University of Art. She chose animation for its unique characteristics and the limitless possibilities it offers to conceptualise and explore impossible worlds. Currently, She teaches at Emily Carr University of Art and Design and has been a regular faculty member in the Animation Department at Tehran University of Art. She has supervised over 70 practical and theoretical MA animation theses and instructed in a wide range of animation workshops for children, youth, and adults. Her expertise and interests span creative writing, narratology, Aristotelian studies, contextual studies, improvisation, art therapy, drawing, short animation, and experimental animation. In addition to her academic pursuits, she has directed seven short animations, written several animation scripts, and contributed to a variety of animation projects.

Jay Russell is a novelist and short story writer. His works include the World Fantasy Award nominated *Brown Harvest*, and three books in the 'Marty Burns' supernatural detective series: *Celestial Dogs, Burning Bright*, and *Greed & Stuff*. His short fiction has appeared in a variety of genre anthologies as well as his own story collection, *Waltzes and Whispers*. In his deceptive, bespectacled Clark Kent guise, he is Dr Russell Schechter, Course Leader for Creative and Professional Writing at St Mary's University Twickenham. He has never knowingly cavorted with a beast of the air under either name.

Rachel Steward is an independent researcher whose work on the image of the blue sky within contemporary visual culture has taken the form of conference papers, talks, publications, a web archive (www.thinkingbluesky.co.uk) and a film contributed to Multiplicity's *Border Devices* convened within the Venice Biennial (2003). In parallel to her research she has worked as a lecturer, public arts programmer and curator including award-winning commissions with Rose Finn-Kelcey (2004) and Droog Design (2004). She also produced and edited *Engaged* Magazine (1994–1998), an experimental arts magazine, copies of which are held in The National Art Library, Victoria and Albert Museum (London) and Goldsmith's Special Collections (London).

Dr James Williamson is a lecturer and programme leader in Media and Communications at the University of Winchester. He is interested in the historical and contemporary imaginary of science and technology across different forms of media, as well as functions of mythic storytelling in utopian, dystopian and apocalyptic narratives. He is co-editor of the special issue 'Future Genders' (2021) for *Visual Resources* and author of the article 'Cybernetic Soundscapes: Re-synthesizing the Electronic Tonalities of Forbidden Planet' (2022) in *Science Fiction Film and Television*, as well multiple forthcoming book chapters on atomic horror and encounters with archetypal others in the science fiction films of the 1950s (including this one).

The Editors

Dr Jon Hackett is a Senior Lecturer in Film and Television at Brunel University and former Head of the Department of Communications, Media and Marketing at St Mary's University. He is the co-author with Dr Mark Duffett, of the University of Chester, of *Scary Monsters: Monstrosity, Masculinity and Popular Music* (Bloomsbury, 2021). His current research focuses on migration in the media, and political cinemas.

Dr Seán Harrington is an academic and lecturer, who currently lives and works in Dublin, Ireland. He has taught cohorts of students from UCD (Ireland), Boston University (USA), Brunel, West London (UK) and St. Mary's University, Twickenham (UK). While he has previously published work on animation and psychoanalytic theory (*The Disney Fetish*, 2015), he has a special love of weird fiction and all things horror.

Damian O'Byrne is a Senior Lecturer in Media and Communications at St Mary's University, Twickenham and is Course Lead for the university's Foundation Year. His background is in graphic design and he teaches across a range of practical modules that focus on magazine design, digital art and photographic manipulation. His research interests have concerned digital media and specifically the role and impact of live television news from a Baudrillarian perspective. He also edits, designs and publishes SBG magazine, an independent magazine about wargaming in J.R.R. Tolkien's Middle-earth. Damian has recently begun to combine his personal passion for Tolkien with his academic career and is planning to embark on a PhD focussing on the practices of Middle-earth fan communities.